# PERSPECTIVES of PSYCHOLOGICAL OPERATIONS (PSYOP) in CONTEMPORARY CONFLICTS

# PERSPECTIVES of PSYCHOLOGICAL OPERATIONS (PSYOP) in CONTEMPORARY CONFLICTS
## Essays in Winning Hearts and Minds

RON SCHLEIFER

Chapters 1–5 and organization of this volume copyright © Ron Schleifer, 2011, 2012, 2013; chapter 6, *Cultural Warfare: Secularization Defense Initiative*, copyright © Benjamin Brown and Ron Schleifer, 2011, 2012, 2013.

The right of Ron Schleifer to be identified as Author of Chapters 1–5, and Benjamin Brown and Ron Schleifer to be identified as Authors of Chapter 6, has been asserted in accordance with the Copyright, Designs and Patents Act 1988.

4 6 8 10 9 7 5 3

*First published 2011 in hardcover, reprinted in paperback 2013, in Great Britain by*
SUSSEX ACADEMIC PRESS
PO Box 139
Eastbourne BN24 9BP

*Distributed in North America by*
SUSSEX ACADEMIC PRESS
ISBS Publisher Services
920 NE 58th Ave #300, Portland, OR 97213, USA

All rights reserved. Except for the quotation of short passages for the purposes of criticism and review, no part of this publication may be reproduced, stored in a retrieval system, or transmitted, in any form or by any means, electronic, mechanical, photocopying, recording or otherwise, without the prior permission of the publisher.

Published with the financial assistance of the School of Communications, Ariel University Center, Israel.

*British Library Cataloguing in Publication Data*
A CIP catalogue record for this book is available from the British Library.

*Library of Congress Cataloging-in-Publication Data*
Shleifer, Ron.
 Perspectives of psychological operations (PSYOP) in contemporary
  conflicts : essays in winning hearts and minds / Ron Schleifer.
   p. cm.
 Includes bibliographical references and index.
 ISBN 978-1-84519-454-3 (h/b : alk. paper)
 ISBN 978-1-84519-585-4 (pbk : alk. paper)
  1. Psychological warfare—Israel. 2. Psychological warfare—Palestine. 3. Al-Aqsa Intifada, 2000– —Psychological aspects.
 4. Terrorism—Prevention—Government policy—Israel. 5. Israel—Foreign public opinion. 6. Palestine—Foreign public opinion. I. Title.
 UB277.I75S344 2011
 355.4'1—dc22                                                              2010046881

Typeset and designed by Sussex Academic Press, Brighton & Eastbourne.

This book is printed on acid-free paper.

# Contents

*Acknowledgments* v

1 The Threat: Radical Islamic Organizations 1

2 The Palestinian PSYOP Campaign against the Neutrals during the Al-Aqsa Intifada 22

3 From Oslo to Jerusalem: Fifteen Years of Palestinian Psychological Warfare against Israel (1993–2008) 39

4 Hasbara, Propaganda, and Israeli Public Diplomacy: A Historical Perspective 70

5 Countering Islamic Terrorism: The Psychological Warfare Perspective 98

6 Cultural Warfare: Secularization Defense Initiative 117

*Notes* 147

*Index* 174

# Acknowledgments

This book is the culmination of thoughts, discussions and writings arising from what was a turbulent decade for Israelis and westerners alike. Dealing with Armageddon-like events could have posed a serious challenge to my sanity, had it not been for my family, friends and colleagues who listened, counter-argued, co-authored and responded to issues that are normally kept within close academic and think-tank circles.

I would like first to thank my wife for following me into the academic desert of PSYOP, and my children, for adding much to my understanding of persuasion; thanks go also to my entire family who demonstrated theoretical and practical psychological know-how, and to my friends whose opinions matter much more than they think.

I am grateful to Dr. Benjamin Brown for agreeing to place our co-authored paper in this book and to Prof. Shmuel Sandler for his assistance regarding the chapter on the origins of Jewish Hasbara. Anthony Grahame, Editorial Director at Sussex Academic Press, has assisted admirably in commenting, editing and producing the book. Last but not least, I am indebted to the Ariel University Center for its financial support.

On a sadder note, I would like to commemorate recently deceased Professor Phil Taylor, a true pioneer in the field of PSYOP, who taught me a great deal.

RON SCHLEIFER
Jerusalem, January 2011

# PERSPECTIVES of PSYCHOLOGICAL OPERATIONS (PSYOP) in CONTEMPORARY CONFLICTS
## Essays in Winning Hearts and Minds

# 1

## The Threat
### The Radical Islamic Organizations

The threat posed by various radical Islamic organizations to the democratic West is perhaps one of the greatest challenges facing the international system at present. In many ways the significance and repercussions of this threat are equal to those of the Cold War, which affected the world for several decades. A more apocalyptic view considers this threat equal to that which caused the downfall of Rome in the fifth century. Like the earlier struggle between the United States and communist Russia, this conflict has an inherently ideological basis. Fuelled by fundamentalist doctrines, radical Islamic groups oppose and reject everything the West stands for. Asserting that Islam is central to both state and society, they seek to impose a strict Shari'a code on the Islamic world, while ultimately aiming to Islamize the entire planet. Here too there are some points of similarity between the current conflict and the earlier one in which both the Soviet Union and the United States, battling for global influence, sought to impose their way of life, communist or democratic, on the international system. Yet here the likeness ends, for radical Muslims, despising the West and spurning all it has to offer, place few if any, and certainly no ethical, limits on what they will do to achieve their ambitions – unlike the two superpowers. If the United States and the Soviet Union were not above fighting wars to further their goals, they did so by proxy, avoiding face-to-face military confrontations: however fierce the ideological battle between the two, neither wanted to spark a nuclear war. By contrast, when it comes to destroying their professed enemy or realizing their aims, the radical Islamic organizations know no such boundaries, and seek, in the name of *jihad*, to wreck maximum havoc on the West. They consider a variety of methods, from the beheading of a Western hostage in Iraq to a wide-scale suicide bombing in

Afghanistan that claimed thousands of victims, to be legitimate, as in their eyes all non-Muslims are equally guilty.[1]

Unlike the Cold War, which was fought between evenly matched superpowers (or so it seemed at the time), this latest conflict is inherently asymmetrical, with the United States and its allies ranged on one side, and several small insurgent and terrorist organizations on the other. Moreover, like the two superpowers (this being the last parallel between the two conflicts), the United States and the fundamentalist Islamists hope to persuade everyone – allies, friends, enemies, their own home audiences – of the righteousness of their cause. The radical Islamic groups, in addition to targeting Western assets, put a great deal of effort into expanding their ideological and working support base, endeavouring to convince various parties in the West and Islamic world that their battle is a just one. One of the principal means they use to win over the public, secure funds, enroll new recruits, and neutralize criticism, is PSYOP, which is short for psychological operation, and refers to actions designed to influence the perceptions, attitudes, and actions of individuals, groups, organizations, and foreign governments.

The radical Islamic groups tend to exploit PSYOP, perhaps the ultimate weapon of the (temporarily) weak, to the full. Unhindered by moral considerations and subject to few if any bureaucratic constraints, they are able to speedily produce and dispatch a range of ingenious messages. In an attempt to disseminate these messages as widely as possible, they operate several hundred websites that describe the inequities of the West, the virtues of Islam, or both. Committed activists as well as fellow travellers scour the world addressing receptive audiences in both Muslim and Western countries, while sympathetic mullahs deliver sermons to the faithful in an effort to convert them to the cause. Well versed in the workings of the Arab media, these organizations exploit channels such as *Al Jezeera* and *Al Arabia* to promulgate their messages. Nor are the Western media, and their insatiable need for dramatic news and images, neglected. On the contrary, many of these groups' activities – most notably the executions of Western hostages and later scenes of Islamist victory in battle – are planned with the media in mind, the aim being to maximize the operations' psychological impact by gaining comprehensive media coverage.

Many of those recruited to fundamentalist Islamic groups are familiar with and even expert in the ways of the West. Some were born in the West, others emigrated there; some are Westerners who converted to Islam, others are members of indigenous minority Muslim communities. Other than providing, thanks to their Western credentials, an effective cover for these organizations' criminal activities,[2] these recruits are also a vital cog in the fundamentalists' PSYOP offensive: for however hostile towards the West they may be, they are able, in the name of the cause, to adopt Western lifestyles, exploit Western democratic values, and take advantage of modern Western technology, all with relative ease. *Au fait* with Western culture, sensitive to its nuances – civic, intellectual, ethical, and linguistic – they exploit and turn this knowledge against the West. Hence, they play on the importance attached in the West to human rights, or use their command of the English, French or German language to recruit students, born in or living in the West, as well as gain access to the Western media. These already widespread practices are bound to multiply as more and more Muslims immigrate to Europe and the United States.

## PSYOP: The Difficulties in Meeting the Threat of Radical Islam

Ever since the 11 September 2001 terrorist attack in the US, the West has been strategically on the defensive. Fearing similar deadly assaults against targets in Europe and the United States, the West has, quite rightly, concentrated its efforts on foiling terrorist acts of carnage. What it has not done, however, is take the war directly to these organizations and wage an effective campaign against them in their own backyard. This is hardly surprising given that the enemy consists mostly of clandestine underground movements and networks. Highly secretive, closed, almost clannish societies, little is known about these groups: their membership lists are a sealed book, their command structures are shrouded in mystery, and most have only the haziest idea as to how they operate. As a result, rather than confront these organizations head on, the West elected to declare war on states known to harbour and/or support them. But even when such wars initially succeeded, as in Afghanistan, they did not lead to the capture of these movements' leaders such as Osama bin Laden, quell the Taliban, prevent them from spreading their ideology among Muslims in either the East or the West, or stop them from carrying out terrorist attacks.

With conventional wars proving largely inadequate to eliminate

THE THREAT

the terrorist threat, PSYOP offers an alternative, and if properly used, effective means for confronting the fundamentalist terrorist challenge. Moreover, as the militant organizations are themselves making extensive use of PSYOP, and with great effect, it is essential that the West launch a PSYOP campaign of its own, if only to neutralize and confound these groups' PSYOP activities. Unfortunately, however, the West has long had an innate moral aversion to PSYOP, which renders any comprehensive and systematic use of it moot. Ponsonby, who wrote on British PSYOP in World War I, put the prevailing liberal view best: "The defilement of the human soul," he said, "is worse than the destruction of the human body."[3]

Within the world's democracies, PSYOP, with its connotations of mind-games and brainwashing, provokes feelings ranging from slight unease to revulsion. Admittedly PSYOP, like any tool, can be used indiscriminately and without regard for moral boundaries. It enables regimes to tailor messages according to their needs, or in more brutal terms, to distort reality at will. These skewed messages are then directed at specific audiences in a highly focused and effective manner. Perceived as a moral outrage, an affront to liberal values, PSYOP has been largely disavowed by Western democracies, at least in times of peace.

Yet, however profound their aversion to PSYOP, once World War II broke out, the Western democracies found themselves forced to adopt PSYOP tactics.[4] Faced with an enemy who fully embraced PSYOP and devoted tremendous resources to launching and prosecuting a PSYOP campaign, they had little choice but to follow suit. This was equally true during the Cold War, in which the West was obliged to counteract the Soviet Union's highly sophisticated PSYOP infrastructures.[5] Yet even then, and even during the Vietnam War, when the Vietcong used PSYOP effectively, the West, specifically the United States, limited its PSYOP activity. It entrusted most PSYOP to the hands of the secret intelligence services, and by doing so undermined its effectiveness. In the early 1980s, with the Cold War once again intensifying, the United States took a greater interest in PSYOP. Ronald Reagan and his advisers admitted that in the fight against Communist Russia – "the Evil Empire" as Reagan called it in clearly PSYOP-ian terms – it would be a grave mistake to abandon the PSYOP battlefield to the enemy. PSYOP was given priority.[6] Caspar Weinberger, the American Defence Secretary, ordered the army to produce a comprehensive

PSYOP policy and to develop on its basis plans of action as well as infrastructures designed to advance the West's practical and ideological goals. But once the Cold War ended, most of the United States' PSYOP-related apparatus was dismantled.[7]

Nevertheless, the seeds were sown, and PSYOP, now conceptually more acceptable, was increasingly integrated into Western strategic thought. Integrated, but still limited in scope; the reason being that for last twenty years the United States has regarded PSYOP largely as a sub-branch of information warfare.[8]

PSYOP may be a morally ambiguous tool with potentially totalitarian overtones, but if the West wants to survive and defeat the Islamic fundamentalist onslaught, it must overcome its honourable, if, under current conditions, misguided misgivings with regard to PSYOP. (Almost) all is fair in love and war, and if the West hopes to quash the extremist Islamic threat and vanquish an enemy whose ultimate goal is to destroy it and all it stands for, then it has little choice but to engage in a PSYOP offensive. Without stooping to the abominations of the militants' PSYOP campaign, the West must mount an effective PSYOP war of its own. Not to do so would mean abandoning the PSYOP field to the Islamic fundamentalists, with all the attendant implications to the progress and outcome of the war.

PSYOP is still, in essence, more moral than conventional military methods: it does not kill or maim, nor does it result in the wholesale destruction of landscape or property. In fact, given that its goal is to bring wars to an end as quickly as possible and so save lives, it can even be considered a moral imperative. Finally, promoting the values of democracy and working towards a democratic world, which is what PSYOP in its battle against radical Islam seeks to do, are highly creditable goals, which are well worth fighting for – as the West must realize.

## The Difficulties of Waging a PSYOP Campaign against Radical Islam

When launching a PSYOP campaign against fundamentalist groups it is important to bear in mind PSYOP's limitations, both in general and as regards radical Islam in particular. By delivering to enemy audiences messages, PSYOP aims, among other things, to stun its adversaries to the point that they are unable to regroup and thus incapable of launching an effective PSYOP counter-offensive of their own. It is not, however, enough simply to deluge the opposition with messages at random; a successful PSYOP offensive depends upon identifying and then adroitly exploiting to maximum effect any and all PSYOP opportunities. And therein lies the difficulty. The most innocuous of events, a collapsed building, a violent domestic incident, or even a run-of-the-mill accident, can be exploited to paint the enemy black, but only if the opportunity is recognized and the information cleverly twisted to the desired effect. Unfortunately, nothing could be easier than to overlook the opportunity afforded by a spate of upset stomachs to accuse the enemy of wholesale poisoning, or to fail to see how a closing of accounts between criminals, ending in a bloody shootout with several victims, can be presented in the press to charge the enemy with viciously pursuing a policy of assassination.

The ability to seize upon and fully exploit such opportunities – one of PSYOP's principal operational tenets – demands the establishment of sizeable PSYOP infrastructures manned by professionals. At the same times these infrastructures must be substantial, they must also maintain tight security, for deploying PSYOP demands extraordinary sensitivity: a single leak can affect the credibility of the messages sent or the integrity of the various channels of communications used, rendering them utterly useless.

Seizing the PSYOP moment also demands detailed and ample anthropological and cultural data on the various audiences targeted. This raises another problem, specific to the battle against radical Islam. Edward Said, in his influential book *Orientalism*, cast doubt on the validity and integrity of the work of Western scholars specializing in Islamic and Arabic affairs. Not only did these scholars, Said claimed, view the Arab/Islamic world from the outside, but they did so through the thoroughly distorted prism of Western imperialism. As a result, he concluded, not only are they ill-equipped to understand, analyse, or evaluate Islam, but they have no right to do so. In the West, his comments fell on receptive ears, wracked as Westerners were with guilt for their countries' imperialist pasts. Anxious to mend their ways, some western scholars went too far in the other direction, and in line with Said's criticisms adopted the Arab/Islamic point of view wholesale. In the United States, with Saudi Arabia funding much of the work done on Arab and Islamic issues, objective research in the field suffered another blow, as most scholars are naturally reluctant to bite the hand that feeds them.[9]

In waging their PSYOP campaign, the radical Islamic organizations have a number of inherent advantages over the West. Utterly convinced of Islam's ultimate victory – it being God's will – they imbue most of their messages directed at Islamic audiences with a sense of teleological inevitability, making it very difficult for the West to counter such triumphalism, let alone garner support among the more moderate Islamic elements. In addition, as small, clandestine organizations seeking to subvert and destroy Western civilization, they have no problem initiating actions such as shooting at soldiers or policemen from within a civilian crowd, thus provoking a violent response and civilian casualties, in order to engineer a PSYOP coup. States, and democratic states in particular, are morally, legally, and institutionally barred from carrying out such operations, and rightly so. Yet even when working within a rigid bureaucratic framework and subject to clear ethical standards, the West may still find the necessary flexibility to instigate equally creative but morally defensible operations.

## Application: Learning from the Past

Though PSYOP has been integrated into the West's war plans since 1991, the West has still much to learn about the subject, especially against radical Islam. Unfortunately the West has muddled through, relying on a process of trial and error, rather than learning from past experience. Had this been done, the United States might not have had to abort its plan to establish the Office for Strategic Influence in October 2001, and a decade later the Office for Strategic Deception in January 2010; these organizations were to coordinate and collate intelligence on terrorist organizations. A leak to the media, probably the result of a bureaucratic power struggle, produced a public outcry, with many turning against the plan simply because of the undemocratic connotations of the department's name: a silly mistake which could easily have been avoided.

In the context of the current battle against radical Islam, specific note should be taken of the part played by PSYOP in the wars waged against Arab and Islamic countries over the past two decades. The American-led campaigns in Afghanistan and Iraq established PSYOP's credentials as a powerful weapon of war. It had an immense impact – at least while the war was raging – on both the enemy soldiers and civilian population; indeed its contribution to the victory on the ground cannot be over-estimated. By contrast, Israel's response to the Palestinian PSYOP campaign, which has been at worst lax, and at best ineffective, proved the dangers of ignoring PSYOP, particularly in low-intensity conflicts, like the one waged by the Western Coalition in Iraq since March 2004. Affluent and technologically sophisticated though the West may be, it will pay a heavy price if fails to devote sufficient time, effort, and resources to waging a comprehensive PSYOP campaign against the Islamic threat.

### Examples from the First Gulf War

The First Gulf War was a watershed in the use of PSYOP in wartime: over 70,000 Iraqi soldiers surrendered to US forces without a fight. These soldiers were found clutching the safe conduct pass composed by the US psychological warfare units and distributed prior to and during the invasion. The Iraqi soldiers, who had taken note of similar safe conduct promises broadcast over the radio, held on to these leaflets, secreting them about their bodies, despite the risk of severe punishment if they had been caught with them in their possession. This surrender by thousands of Iraqi soldiers saved the lives of countless soldiers, both American and Iraqi; all at the price of several tons of paper and a few radio transmitters. In short, PSYOP helped win the war, and the United States pulled off one of PSYOP's primary wartime goals: persuading enemy forces to stop fighting.

Throughout the war, the United States made use of black PSYOP techniques, for example by setting up clandestine radio stations. The programs broadcast on The Voice of Free Baghdad, ostensibly run by the Iraqi opposition, were directed at the Iraqi population and designed to undermine local support for Saddam and the war. The United States also managed to dupe the Iraqis regarding the principal point of attack. Prior to the invasion, the United States made sure that leaflets entitled "Marine (Shock) Wave" fell into Iraqi hands. But rather than indicating the precise point of the impending attack, the army was content to pack the leaflets with an image likening the marines to an all-engulfing wave, sweeping away all in its path, and leave the Iraqis to draw their own conclusions. They did so, inferring that the main attack would come from the sea. The leaflet was distributed by local fishermen who sent it inside some 20,000 plastic bottles; the leaflets persuaded the Iraqi high command to concentrate their forces along the seafront in preparation for the American assault.

The First Gulf War marked a turning point in the relationship between the US armed forces and the media, thanks in part to the demands posed by the PSYOP campaign waged at the time. Government manipulation of the press, both prior to and during the war, was intense, as the US armed forces sought to exploit the media to further their psychological crusade against the enemy. This in itself was nothing new; what was new was the unprecedented scale of media manipulation. When Saddam made good on his threat to flood the Persian Gulf with crude oil, the United States seized on what it rightly saw as a superlative PSYOP opportunity. Giving the media access to the area, it allowed reporters to publish pictures of the devastating impact Saddam's action had on the local wildlife and ecosystem. Photographs of a dying cormorant desperately trying to flap its oil-covered wings affected millions throughout the world.[10] It was a clever piece of PSYOP given that in today's world the pitiful image of a frantically struggling bird can often strike a stronger emotional chord than endless descriptions of famine-racked countries, to which the public has, unfortunately, become somewhat inured. This and other similarly compelling pictures reflected the growing recognition that in terms of winning the war, images of the war, stage-managed or not, were no less important than the battles fought on the ground.

## The War in Afghanistan

The war in Afghanistan broke out in October 2001 in the aftermath of September 11, and in some ways ended in January 2003 with the beginning of the Second Gulf War, only to be recommenced in 2008. The PSYOP campaign waged by the United States during the war was notable for three things. First, for the way in which it overcame the hurdles generated not only by the vast cultural gulf between the United States and the local population, but also for the facts that very few of the largely rural Afghani population spoke any English, and that not all Afghans could read or write the local Deri or Pashtu dialects. Second, for the emphasis given to consolidation PSYOP as part of a concerted effort to address the problem of what happens once the fighting on the ground stops. Consolidation

PSYOP, which deals with the political, social, and economic aftermath of war, is rooted in the American view of PSYOP as a comprehensive, all-inclusive form of warfare, which aims at promoting the country's strategic-political goals as well as its military objectives.[11] Third, for the way the army, in line with the policies adopted during the First Gulf War, took great care to inform both the American public and Capitol Hill of its various PSYOP endeavours.[12] Though aware that such exposure might undermine the effectiveness of its PSYOP messages, the United States administration nevertheless decided that it was a risk well worth taking in order to secure public and political backing for the war. In the event, these disclosures had little effect on the PSYOP campaign, possibly because, as noted earlier, most of the Afghani population were unaware of events in the United States.

The PSYOP offensive targeted two audiences: the civilian population and the anti-American forces, primarily the Taliban and its supporters. The campaign was launched even before the fighting began: that is, from the moment the Taliban and its supporters were identified as the enemy. Given the underdeveloped communications infrastructure in Afghanistan, the United States was forced to fall back on modes of communication dating back to World Wars I and II in order to relay its PSYOP messages. Indeed, the humble leaflet proved to be the most effective, and sometimes the only possible method of circulating PSYOP messages in Afghanistan. Having experimented with leaflets of all shapes and sizes, the United States finally settled on a piece of paper the size of a banknote and featuring an illustration or caricature, combined with a short written message. The use of graphics was designed to overcome the problem of widespread illiteracy. Radio broadcasts proved to be another cheap and useful method for surmounting this problem; and as many of the more remote and poverty-stricken regions of Afghanistan did not have a radio set, the Americans dropped thousands of cheap transistor radios in the countryside. Hoping to attract as many youthful listeners as possible, they broadcast local popular music, banned under the Taliban regime. The fact that the ethnically diverse Afghanis speak several languages was another difficulty faced by United States PSYOP units. Simply illustrated messages were one solution to the problem; another was to write the messages in both Pashtu and Deri and hope for the best.

An analysis of the content of the American PSYOP messages in Afghanistan reveals that great care was taken to ensure their credibility. Given most peoples' instinctive inclination to treat with suspicion and reject any messages originating from the enemy, PSYOP messages must have a high degree of integrity and veracity if they are to be assimilated. Moreover, as PSYOP operatives have learnt to their dismay, it takes just one evidently false message to destroy the credibility of a whole channel of communication, if not an entire PSYOP campaign. Hence, in Afghanistan the Americans chose to propagate their messages through radio news bulletins and news commentary. The news is generally regarded, rightly or wrongly, as a highly credible source of information; in the Afghani case this credibility was enhanced by the fact that these broadcasts came from seemingly independent Islamic sources.

The successful American PSYOP campaign in Afghanistan, in short, contains a number of lessons on how to wage PSYOP campaigns in similarly undeveloped regions.

## Iraq following the Second Gulf War (2003–2004)

While planning the war in Iraq, the United States focused on winning the war on the ground, but devoted almost no attention to the situation that would follow the ground war. This, among other things, affected the nature of its PSYOP campaign, which was initially aimed at helping the coalition forces achieve victory. As a result, once the fighting subsided a political vacuum emerged, which local insurgents were quick to exploit, thus endangering the success that had been achieved. The insurgents also took the lead in the emerging PSYOP battle. At the outset, their campaign was based largely on rumour-mongering and graffiti, but as local dissatisfaction with the occupation grew, the insurgents became increasingly sophisticated, feeding skewed information to regional television and radio stations. Qatar's Al Jezeera news network in particular was frequently favoured with choice bits of PSYOP information thanks to its close working relationship with the Western media.

Having realized that without a massive consolidation PSYOP offensive, local opposition to the occupation would skyrocket, the United States set about rectifying its unfortunate oversight. However, mounting a consolidation campaign in post-war Iraq was, as the Americans soon appreciated, a far from easy task. They were faced with a population that for more than three decades had been indoctrinated by a brutal dictatorship; the local media were far from sympathetic; and anarchy and institutionalized terrorism raged throughout the country. Given that for years the local population had been subjected to anti-American propaganda under Saddam, while its experiences under occupation – for example, the coalition's failure to restore essential services such water, electricity, and communications quickly enough – failed to inspire confidence in the US, it is hardly surprising that the insurgents' vilification of the United States as "The Great Satan" or "New Crusaders" fell on increasingly receptive ears. The widespread publicity given to the Abu Ghraib affair and the sadistic abuse of Iraqi prisoners by American soldiers played right into the insurgents' hands, making it doubly difficult for the United States to shake off its negative image. Add to this the failure to find any Weapons of Mass Destruction, whose alleged existence was the reason the coalition claimed it went to war, and it was almost inevitable that the United States and its allies' reputation among the locals would plum new depths.

The initial failure to capture Saddam Hussein also hindered the American PSYOP campaign, with many an Iraqi fearing the dictator's political resurrection and with it the reinstitution of a vengeful Ba'ath regime. Following the capture of Saddam things became easier, as the Iraqis realized that a point of no return had been reached in Iraqi history and that the coalition was truly committed to ridding Iraq of Saddam. This allowed the United States to disseminate its PSYOP messages with greater success. In an attempt to neutralize Al Jezeera and its like – often seen as little more than spokespersons, even if unwittingly so, for the insurgents – the United States established its own media networks, most famously the satellite television station Al Hurra (The Free). Al Hurra, which was said to have cost the US government $62 million in its first year of operation,[13] like the daily newspaper *Baghdad Times* and Radio Sawa (Friendship), a youth radio station broadcasting pop music punctuated by news bulletins, was set up specifically to disseminate US PSYOP messages. The American victory in Faluja, a Shi'a stronghold and thorn in the side of the occupation forces, proved grist to the PSYOP mill and was exploited to demonize the insurgents. Having captured the town, the United States displayed the cages in which the insurgents had held kidnapped hostages as well as the swords allegedly used by the insurgents to behead them. These images were accompanied by fulsome reports of the locals' immense gratitude towards the Americans for having freed them from the clutches of a Taliban-like regime.

## PSYOP and Technology in the Twenty-First Century

The means of delivering PSYOP messages is often no less important than the content of the messages themselves. Indeed, the principal advance in PSYOP over the last hundred years has been the rapid and unprecedented development in the methods of disseminating messages in time of war. If in World War I the newspaper was the chief means of conveying PSYOP messages, by World War II it had been replaced by radio and film newsreels. During the Vietnam War (1961–1973), most PSYOP messages were channelled through television. The war in Lebanon (1982) saw the beginning of the satellite era. Then, in rapid succession came the fax machine (the First Intifada, 1987–1991); the satellite phone (the First Gulf War, 1991); the internet (Kosovo, 1999); and finally satellite video, which made its PSYOP debut during the Second Gulf War (2003).

The internet, together with satellite radio and video, produced an explosion of information. Messages can now be sent to and from a wide range of wireless devices, including beepers, cellular phones, Blackberries, and iPods. Then there is the internet, which allows the almost instantaneous dissemination of stupendous amounts of information around the world. The advent of the satellite phone with integrated digital camera created a one-man mobile news enterprise, signalling the death of government censorship.

These developments, which have already had a considerable impact on PSYOP, will have a still greater effect in the twenty-first century. The proliferation of cheap, readily available communication technologies will make it much easier for small groups and individuals to get their message out to millions across the world. The ubiquitous internet will prove to be, indeed already is, a particularly useful PSYOP tool, used to send vast numbers of e-mails to people all over the world, and to post websites – often untraceable – which can be accessed by whomever, wherever. Seizing control of and subverting channels of communication, for example by hacking into personal computers or broadcasting on existing radio and television frequencies, will become increasingly easy. Computer viruses that disrupt or replace messages and redirect browsers to specific websites are but one tool in the arsenal of cyberwarfare. Finally, satellite technology, and especially smart phones, offer immediate access to information from almost anywhere. All this will work to the advantage not only of state-driven PSYOP, with the technically sophisticated West exploiting its vast resources to reach audiences worldwide, but also of non-state organizations such as terrorist groups, whose ability to manipulate and disseminate tendentious material will continue to increase.

In this context it is essential not to fall into the common trap of assuming that radical Islamic organizations are little more than a horde of illiterate, primitive peasants. Many Islamic militants, a large number of whom are science graduates, alumni of Western universities and polytechnics, are extremely savvy technologically. According to American sources, in the final stages of the war in Afghanistan, Bin Laden's troops stopped using satellite phones, well aware that these could be tapped in order to pinpoint their (and Bin Laden's) location. One of the reasons the organization's activists in the West use internet cafes in order to stay in touch with one another is that it is virtually impossible to track down the people who send messages from these cafes. Not only are these groups technologically literate, but they also have access to a variety of communication devices – their own, as well as those belonging to others – all of which can be, and are, harnessed to advance their PSYOP campaign. And what they lack in technological resources they more than make up for through original and creative manipulation of the PSYOP medium.

With PSYOP emerging as an increasingly important element of low-intensity wars, the fundamentalist groups will exploit both current and newly developed communication systems to disseminate their messages to the world's Islamic communities, an area in which they clearly have a head start over the West. Nor will they neglect old fashioned modes of communication such as leaflets, graffiti, megaphones, and rumour campaigns, all of which have proved remarkably effective in Islamic countries. The radicals have an additional advantage over the West in that they are a part of the community they seek to convert. This means that they are familiar with the working of that society, and also that credibility is less of a problem. They will exploit Western technology to launch a vigorous PSYOP campaign against the West, disseminating prodigious amounts of material, most of it difficult if not impossible to trace, and this will allow them to operate with increasing impunity. Yet the West has the technological resources, if not to silence the militants' various channels of communications, then at least to smother them with a massive PSYOP offensive of its own.

## The Cultural Dimension of PSYOP: Anthropological Surveys and Analysis

Nowhere has the expression "know thy enemy" more resonance than in the field of PSYOP. Following the events of September 11, the United States concluded that one of the reasons it had failed to detect and prevent the terrorist attack was that neither its intelligence community nor freelance academics were sufficiently conversant with the myriad nuances of Islamic culture and creeds. Similarly, there is little doubt that many of the difficulties faced by the coalition in Iraq in the aftermath of the Second Gulf War were due to its inability to grasp the complexities of Iraqi society. As a result, and in an attempt to make good its failings, the United States began to devote an increasing amount of time and effort to gathering, collating, and analysing information on the social, cultural, anthropological, and religious make-up of hostile and potentially hostile – mainly, it had to be said, Islamic – states and entities.

Conventional military intelligence focuses on such questions as the enemy's location and military capacity. It contributes to battle plans and strategies, and aims to be sure that the best and most appropriate weapons are available where needed. Cultural intelligence, by contrast, particularly when PSYOP-orientated, seeks to build up a complete anthropological picture of the state, organization, or society targeted. Its aim is to pinpoint which psychological and/or cultural buttons to push, and when, during the PSYOP campaign. This includes identifying the cultural codes that can be used to persuade the enemy soldiers to surrender, detecting signs of war-weariness, both in society at large and among the fighting forces, as well as noting the emergence of divisions within the enemy camp, whether between senior and junior officers or between the army and the civilian population.

If, as in the case of the war on terror, the PSYOP goal is a strategic one – for instance, disrupting relations between states – it is essential to remember that the target audience is not necessarily monolithic. This is certainly true of the Islamic world, where

Malaysia is different from Afghanistan, and both differ from Iraq. In terms of gathering intelligence, each state must be tackled individually and an effort must be made to determine its discrete cultural characteristics and idiosyncrasies. It is equally important to bear in mind that most Islamic states or societies boast several communities, some moderate and others radical, and that the former, no less important than the latter, must also be probed, and their cultural proclivities determined. When there is sufficient data on these admittedly complex phenomena, and good coordination with the political powers that be, the chances of developing a successful and cost-effective PSYOP campaign become significantly greater.

Gathering and then drawing on cultural and anthropological information in order to formulate effective PSYOP messages is an immense and complex undertaking. Cultural intelligence demands a perfect command of the *lingua franca*; an appreciation of local customs, religions, and interpersonal codes of communication; more than a passing acquaintance with native myths, legends, and folklore; and above all, an understanding of exactly how the locals think. Given the vast amount of variegated material involved, there is clearly a need for a network of PSYOP research centres focusing on the collection and analysis of anthropological, cultural, psychological, and linguistic data.[14] Special PSYOP units, manned by operatives fluent in the local languages and dialects and conversant with the local cultures, are of course mandatory. Finally, even though most cultural intelligence can be gleaned from overt, easily accessible sources, such as books, newspapers, and interviews with members of the target groups, in order to fill in any gaps, an international information sharing network ought be institutionalized among Western intelligence services.

The effort to obtain and exploit anthropological intelligence will prove doubly effective if conducted on a global scale, with countries sharing research methods and pooling their findings. States which border Islamic nations, or which have direct dealings with them, or whose population includes Muslim minorities, have a lot to contribute in this respect, boasting several advantages in terms of gathering and assessing PSYOP intelligence. The knowledge and experience these countries have amassed thanks to their dealings with Islamic society – Israel, for example, has faced a grave Arab-Islamic threat for over sixty years – could do much to further the global PSYOP intelligence campaign against Islamic countries.

## Consolidation PSYOP

Cultural intelligence is key to the success of consolidation PSYOP. Consolidation PSYOP aims at winning over the conquered population's hearts and minds. It seeks to accustom them to the fact of occupation. It endeavours to convince them that the occupying force has their best interests in mind, and will do all that it can to accommodate their needs and certainly do nothing to harm them. Finally, at least in some cases, such as that of Iraq in 2005, it hopes to impress upon the conquered population that the occupation is temporary. In the aftermath of the Second Gulf War, the successful insurgent attacks against foreigners, mostly Americans, in Iraq, only served to underline the vital importance of consolidation PSYOP.[15]

As the fighting in Iraq died down, it became clear that in planning a war of conquest, even when the occupation is intended to be temporary, it is not enough simply to assemble a well-trained and well-equipped fighting force capable of crushing the enemy. There is also a need to prepare for the day after, and not only in terms of setting up physical, political, and economic infrastructures, but also in terms of initiating a massive intelligence operation geared towards consolidation PSYOP. This means mapping the target society, taking particular note of the cultural and religious differences within that society and between it and the West. Such information, combined with data gleaned from conventional intelligence sources, enables the occupying force to pursue policies attractive to the civilian population, and to adopt strategies and tactics designed to win their hearts and mind.

## Computerized PSYOP Systems

There has been some experimentation in recent years with computerized PSYOP systems.[16] Fed with detailed scenarios and anthropological, cultural, and linguistic data, these systems have the potential to generate a rapid stream of PSYOP messages in times of war when PSYOP units must react quickly to unexpected changes in the field. In theory these systems could produce leaflets phrased in terrifying language to induce fear, or alternatively, generate safe-conduct passes which persuade the enemy to surrender. Computerized systems could also facilitate the process of distributing PSYOP messages. By analyzing technical data on conditions on the ground as well as the state of the weather, they could recommend whether it would be best to circulate leaflets by plane, missile, or hot air balloon, for example.

There is however one problem: cultures are by their very nature intricate, multifaceted things, replete with subtle nuances of thought and behaviour. Composing effective PSYOP messages that home in on the enemy's cultural sensibilities demands careful attention in choosing graphics and concepts that will appeal directly to the target audience's proclivities. No computer program can do this; it can only be done by experienced PSYOP operatives, expert in the language and culture of the target society. The goals that PSYOP messages hope to achieve are extremely ambitious: they often seek to induce a person to act against his or her habits or declared purpose, be it to surrender, to change their beliefs, to take part in a battle, or to die for a cause. PSYOP endeavours to convince indoctrinated Islamic militants to surrender, defect, or despair of achieving their goals. No machine, however sophisticated, is capable of producing such messages. Clearly with regard to PSYOP there is no substitute for the human touch.

But this does not mean that computers need be relegated to the sidelines of the PSYOP game. They can offer a quick interim solution to the problem of producing PSYOP messages in a crisis or in the heat of the battle. Hence, time can be gained until the PSYOP units arrive on the scene and take over. Naturally there is the danger that purely computer-generated messages could undermine the whole PSYOP effort, with a wrong word in the wrong place or a simple misspelling producing widespread ridicule or resentment. But this risk must be weighed against the advantage of firing off in quick succession messages that could prevent the situation from deteriorating. And once things calm down the computer systems can offer a variety of PSYOP options, which the human PSYOP operatives can accept, reject, or refine as needed, thus saving both time and manpower.

## Tackling the Suicide Bombers

PSYOP is predicated on the assumption that people are above all animated by the will to live. It is true that in battle soldiers may be fired by patriotic feelings and suppress their natural instinct to survive, and thus be willing to die for their country or for their comrades in arms. Yet under certain conditions, carefully written PSYOP messages which dwell incessantly on the enemy's – or more precisely the individual soldier's – desperate situation, can prove remarkably effective and can render enemy soldiers unwilling to make this type of sacrifice. Messages highlighting the miserable state of the enemy's fighting force, its generals' ineptitude, its officers' duplicity, and the inadequate medical facilities, even if not wholly believed, are enough to sow the seeds of doubt, which in turn reactivate the target's human survival instincts. Once convinced that their side is bound to lose, that they are fighting for a lost cause, most soldiers, even the most fanatical, will, as history has demonstrated, conclude that it is better end up in POW camp, alive, than dead in body bag.[17]

None of this is relevant, however, when it comes to suicide bombers. Islamic suicide bombers deliberately seek out death in the belief that they will be lavishly, indeed lasciviously rewarded for their deeds in heaven.[18] Literally embracing death, the suicide bomber is immune to the underlying psychological assumptions of PSYOP. Various countermeasures, like burying suicide bombers in pigskin, which according to Qu'ranic teachings would prevent them from entering heaven and enjoying the ministrations of seventy virgins, proved useless once the fundamentalist mullahs – mostly the very same mullahs who had sent these suicide bombers on their deadly missions in the first place – issued a ruling nullifying the pigskin's alleged prejudicial effect. Yet, if the phenomenon has so far proved a hard nut to crack, there is no need to despair. The practice of suicide bombing is rooted in a particular, indeed perverse (as many Muslims would admit) reading of Islam,[19] and the key to tackling this threat lies in carefully studying the numerous interpretations of the Qu'ran and the Hadith as well as sundry clerical rulings; recruiting moderate clerics to the cause; and launching a PSYOP offensive that emphasizes the traditional reading of Islam, which celebrates life, and refutes the fundamentalist position and its sanctification of death.

## Black PSYOP

Black PSYOP is perhaps one of the most valuable counter-fundamentalist and counter-terrorist weapons in the West's arsenal.[20]

Black PSYOP consists primarily of disseminating messages that seem to come from the enemy, including radio broadcasts from stations ostensibly run by the enemy, and leaflets purportedly written by the opposition forces.[21] These sham communications are used to air spurious – and sometime not so spurious – accusations and/or information, the aim being to confuse, confound, or embarrass the other side. Currently, however, thanks to the efforts of amateur radio buffs who have set out to identify and expose all fraudulent radio broadcasts, establishing bogus radio stations would appear to be a non-starter.[22]

But there are other options: leaflets, for example. During both Intifadas the Palestinians accused Israel of distributing counterfeit fliers which were purportedly the work of Palestinian organizations. The internet is also an ideal platform from which to launch a black PSYOP campaign: bogus internet sites and false identities, both notoriously difficult to trace, can be used to spread disinformation in order to destabilize and subvert radical Islamic groups. Nor can the West ignore Islamic organizations' own black PSYOP pursuits, particularly their successful attempts to mobilize support by such means.[23] Though countering the extremists' black PSYOP manoeuvres is without doubt a difficult and costly proposition, it is well worth the effort, if only to undermine the fundamentalists' so far so successful efforts to expand their support base within the Muslim community.

## Conclusions

In its battle against radical Islamic organizations, the West must recognize PSYOP's unique value as strategic weapon and fully exploit it. In order to reap the full benefits of PSYOP in the current war, the West must invest heavily in the appropriate PSYOP infrastructures: establish research centres specializing in cultural, ethnic, and religious affairs, with particular emphasis on social codes of behaviour; gather teams of experts in fundamentalist lore and radical ideologies; and set up PSYOP training centres to school operatives in both the technical and the human aspects of PSYOP. It must then use these resources to target as many audiences as possible – radical and moderate – in the Islamic world, as well as in Europe's own Muslim communities. It must subject them to a constant barrage of messages, exploiting every possible mode of communication from the humble leaflet to the most sophisticated technological devices. Particular attention should be paid to the education systems, where an effort must be made to highlight the advantages of democracies, both on a general level (e.g., "Have you ever seen democracies declare war against one another?") and on a practical or personal level (e.g., "There are no poor democracies!"). And the West must do all this while at the same time neutralizing the messages of radical Islam: its consecration of suicide and its sanctification of indiscriminate terrorist attacks. No less importantly, it must bring to an end these organizations' successful assault on and subversion of the media, an achievement which as increased their confidence and reinforced their belief in their ultimate victory, making them more fanatical still.

PSYOP, whether playing a lead or supportive role, is crucial to winning wars of any type. Moreover, it is not only a highly effective mode of warfare, but it is also much more moral than its conventional equivalent: PSYOP may play mind-games, but conventional wars kill outright. As the era of the traditional conventional war seems to be ending, with low-intensity wars set to become the dominant form of warfare in the twenty-first century, and global terrorism the primary threat to peace, PSYOP's importance will only increase. Indeed, the present battle against fundamentalist Islam, which embodies both trends, cries out for a full-blown and worldwide PSYOP offensive by the West.

# 2

# The Palestinian PSYOP Campaign against the Neutrals during the Al-Aqsa Intifada

The Israeli retreat from the Gaza Strip in August 2005 marked the end of the Second Palestinian Intifada (the Al-Aqsa Intifada).[1] From the Palestinian point of view, the Israeli withdrawal was the Intifada's most significant achievement. Having forced Israel into handing over the territory it had conquered from Egypt back in 1967, the Palestinians had freed the Gaza Strip from Israeli occupation, a situation that would enable the Palestinians to establish an independent state and later on to develop an independent Hamas entity.

The achievement is all the more remarkable given the prodigious disparity in power between the two rival sides. A regional power, Israel has at its disposal advanced weapon systems and a sizable army. The Palestinians lacked proper military countermeasures, and therefore resorted to psychological operations (PSYOP). By making PSYOP an integral part of the Intifada and Intifada-related activities, both military and non-military, the Palestinians were able to score numerous points in the ongoing battle between the two rival sides, achieving several victories in the process.[2]

The few studies of the Second Intifada published to date focus primarily on the conflict's conventional military and political aspects; the question of PSYOP has been largely neglected.[3] This chapter seeks to describe the Palestinian PSYOP campaign in detail, focusing particularly on the PSYOP campaign aimed at the neutral audience. After identifying this campaign's objectives, I will examine its themes and the techniques and channels of communication used to advance them.

## Psychological Warfare

Psychological warfare has proved an increasingly important factor in modern armed conflicts of all types, from conventional wars between states to limited conflicts between unequal combatants.[4] Psychological warfare seeks victory by affecting three key audiences: domestic, neutral, and enemy.[5] It aims to persuade them by both military and non-violent means to support or oppose a particular course of action or policy.[6] PSYOP has been developed into a highly effective weapon in limited, long-term conflicts, in which one side or both are unable to fully exploit their conventional military resources. But it is in asymmetric conflicts that PSYOP truly comes into its own. In such conflicts, it is paradoxically the weaker side that generally determines the nature of the conflict, that is, whether it will develop into a guerrilla war or be characterized by terrorist activities. Either way, the aim is to prevent the more powerful adversary from making full use of its military resources. The weaker side may launch PSYOP campaigns to manipulate not only the enemy's psyche and perceptions, but also that of its own home audience and of various influential neutrals.

In cases where the weaker of the two combatants is a new, emerging liberation movement, it will have several crucial advantages over its adversary. First, because of its smaller military capability, it will tend to give priority to PSYOP-related activities, which are much cheaper than conventional military weaponry, requiring merely cameras, computers, and the like rather than expensive fighter airplanes and tanks. As there is a direct positive correlation between the size of the role allotted to PSYOP within the overall military policy and its ultimate effectiveness, such entities can expect good results in their PSYOP campaigns. Moreover, most sovereign states are inclined to rely on their superior military capabilities to bring about victory; consequently, they are prone to allot PSYOP a minimal role within their defence policy.[7] Thus, the apparently inferior side can actually be superior to its opponent in terms of psychological operations.

Second, unlike states, smaller entities are not burdened by unwieldy bureaucracies, which tend to react slowly to events and to adhere to stock responses. A small organization can respond quickly, creatively, and flexibly to events, all of which abilities are crucial to the waging of a successful PSYOP campaign.[8]

PSYOP has also proved effective in cases where the other combatant had no coherent or single, united ideology, the lack of which made it easier to sap its national vigour and mental fortitude. PSYOP is even more effective in cases where the adversary has no experienced professional PSYOP organization and so is unable to mount a powerful PSYOP counterattack of its own. Both circumstances held true in the Israeli–Palestinian conflict, giving the latter a distinct PSYOP advantage over the former.[9]

## Palestinian Objectives

Seeking to garner as much international support as possible, the Palestinians regarded neutral audiences as key PSYOP targets. They tailored their activities so as to appeal to and win over Western minds, to the point of Abu Mazen acknowledging the sufferings of the Jewish people, an unprecedented statement by a Palestinian leader.[10]

The Palestinian PSYOP campaign's objectives vis-à-vis neutral audiences in both the United States and Western Europe were twofold. First and foremost, the Palestinians sought to obtain international support for their strategic political aim of establishing an independent state. Second, they strove to delegitimize Israel, presenting it as a rogue state bent on breaking international law. They sought to create an image of Israel as a state that treats the Palestinians with excessive and bestial cruelty. In a later phase, their goals would come to fruition through the Goldstone Report on the Gaza War of 2009.[11]

## Themes

In the course of the Second Intifada, neutral audiences were subjected to two principal PSYOP themes by the Palestinians: demonization and blackening the adversary's name. Though similar in nature, these two themes are not identical. Messages that depict the enemy in as negative a light as possible are designed to achieve specific political objectives, such as delegitimization.[12] Demonization themes go a step further and aim at dehumanizing the enemy, rendering him a diabolical manifestation of evil incarnate.[13]

## Blackening the Enemy's Name

The Palestinians accused Israel of systematically violating both human rights and international law. Israel was characterized as a Nazi state while a personal character attack followed against its prime minister, Ariel Sharon.[14] The international media were supplied with stories and images highlighting Israeli abuses and Palestinian sufferings.[15] They accused Israel of using a blend of excessive force and heavy weapons – live ammunition, tanks, missiles, and even nuclear weapons[16] – in order to disperse demonstrations, which, they emphasized, resulted in heavy civilian casualties.[17] Palestinians described Israel's tactic of destroying houses as a humanitarian catastrophe that left many Palestinian families without shelter, in addition to having lost their possessions.

Furthermore, the Palestinians presented Israel's policy of targeted assassinations as indiscriminate murder of innocents. It was accomplished by accentuating those attacks that also claimed civilian casualties, especially when those instances resulted in the death of non-combatants. Concurrently, the Palestinians took care to suppress the identity of the intended targets of the operations: high-ranking operatives in terrorist organizations.

## Violating Human Rights and International Law

By exploiting the theme of "human rights", the Palestinians hoped to procure a new, international anti-Israeli front. Accordingly, they repeated Israel's various infringements of human rights, genuine and fictitious, producing a series of unfavourable reports on the issue.[18] They took advantage of the presence of the numerous UN teams and human rights organizations, such as Amnesty International, in order to make these abuses public. Inasmuch as the Israeli Foreign Office was ill-equipped to deal with the flood of queries directed to it on the subject and refute the accusations, Israel, in this instance, played directly into Palestinian hands.

Two major issues in connection with the Palestinian PSYOP offensive related to human rights were the military checkpoints and the security fence.

## Military Checkpoints

A constant point of friction between the Israel Defense Forces (IDF) and the local Palestinian population, the numerous IDF roadblocks scattered throughout the West Bank and Gaza Strip became one of the most controversial symbols of the Israeli presence. With the outbreak of the Intifada, not only did the number of checkpoints increase tenfold, but the Israelis also put into place a series of increasingly rigorous security measures. As a result, Palestinians were forced to undergo exhaustive and humiliating search procedures and were frequently frisked in order to ensure that they had not secreted any weapons or explosive belts about their persons, as had been known to occur. Many Palestinians were also taken aside, seemingly at random, and questioned at length. A time-consuming and arduous process, it resulted in long and fatiguing lines at these checkpoints.[19] The Palestinians drew the attention of the international press and public opinion to the daily misery caused to the average Palestinian by these checkpoints, misery compounded by the occasionally deplorable behaviour of some of the soldiers manning them. All of this, needless to say, seriously damaged Israel's international image.[20]

### The Security Fence

In the summer of 2002, with the number of terrorist attacks within the Green Line on the rise, Israel decided to build a security fence separating pre-1967 Israel from the West Bank. Denouncing the decision, the Palestinians set about stopping the fence's construction, launching an international campaign designed to convince neutral audiences that in building the fence, Israel was in breach of both human rights and international law. By labelling the fence the Apartheid Wall, they attempted to establish a direct link between the fence and Apartheid South Africa. They further argued that the fence was built on illegally appropriated Palestinian land and that it ruined many Palestinian families economically by separating them from their farmland and main source of revenue. They pointed out that the fence cut the Palestinians off from their places of work, schools and universities, as well as from local medical and municipal services. Finally, the fence split families and friends apart.[21]

The Palestinians distributed photographs of the security fence, almost exclusively of those sections built of concrete, in order to play up its prison-like aura.[22] The Palestinians also raised the matter with the International Court in The Hague, though less in the hope of putting an end to the fence's construction than of condemning Israel in the public arena. They similarly sought to exploit the fact that in some European countries it is legal to prosecute war criminals from other countries, and thus to indict Israeli soldiers and politicians for war crimes; this effort was mostly for publicity effect.[23]

## Damning Israel as a Nazi State

As during the First Intifada, the Palestinians deployed in their PSYOP campaign various historically loaded expressions, such as genocide, ethnic cleansing, and mass liquidation. Their aim was to establish a link in the public's mind between Israel and Nazi Germany. They likewise offered a pseudo-psychological explanation for Israel's behaviour in the occupied territories, claiming that the Jews were attempting to efface the humiliations they had suffered at the hand of their Nazi tormentors by re-enacting that experience, this time in the role of the oppressor.[24]

## Personal Attacks on Israel's Prime Minister Ariel Sharon

Throughout the Second Intifada, the Palestinians used every means at their disposal – speeches, film, texts, websites and caricatures – to traduce and discredit Israel's prime minister, casting Sharon as a monster who devoted his entire political and military career to persecuting the Palestinian nation.[25] Sharon was to become an icon of evil, and naturally the image of all Israelis was damaged by guilt through association.

## Denying Responsibility

Another highly effective PSYOP theme took the form of the Palestinians consistently denying any responsibility for events, the aim being to deflect all criticisms of the Palestinian Authority or militant groups. Accordingly, in response to virtually every suicide bombing, the Palestinian administration responded by issuing a statement condemning the killings, but sympathizing with the motives behind them. After all, the bombers were driven to these desperate deeds by the iniquities and crimes of the Israeli occupation. For example:

> We maintain our strong condemnation of any attacks that target Israeli civilians, especially the latest attack in West Jerusalem. By the same token, we vehemently condemn the massacres that the

Israeli occupation army has been committing in the past fourteen days against the Palestinian civilians and refugees in Nablus City, the Jenin refugee camp, as well as the Church of Nativity in Bethlehem, and other Palestinian areas.[26]

By implying that Israel was ultimately to blame for the bombings, and refocusing attention on the plight of the Palestinian people, the Palestinian Authority managed to thwart Israel's own attempts to appeal to international public opinion. In this way, Israel's attempt to highlight the horrific trauma produced by the suicide bombings went virtually unnoticed by the international media.

## Demonizing the Enemy

This is a long-established PSYOP theme, which, by making use of atrocity propaganda, seeks to divest the opponent of any semblance of humanity. The Al-Aqsa Intifada was much more violent than its predecessor, the First Intifada; it resulted in numerous deaths and injuries, and thus offered many opportunities for the pursuit of atrocity propaganda.[27] By goading the Israelis through guerrilla warfare tactics and suicide bombings into fierce military action – into the use of live ammunition into violent mass demonstrations or targeted assassinations – the Palestinians were able to present the Israelis as predatory, cold-blooded monsters.

Seizing on the momentum, Palestinian spokespersons accused Israel of kidnapping Palestinian children and selling their organs on the black market;[28] distributing poisoned sweets to children; and of sending female soldiers to seduce and infect Palestinians with the AIDS virus.[29] Among the better-known Palestinian efforts in this arena were the 'Jenin Massacre' and the 'Siege of the Church of the Nativity'.

## The Jenin Massacre

In April 2002 Israel launched Operation Defensive Shield, the goal being to destroy domestic terrorists' infrastructure. During the operation, Israel was able to enter and temporarily take control of most of the towns in the West Bank with relative ease. The refugee camps, however, proved a more difficult proposition, made doubly so when Arafat, for PSYOP effect, encouraged the inhabitants of Jenin to fight to the death rather than surrender to the Israeli forces.

The battle for Jenin was thus transformed into a heroic tale, in which ill-equipped Palestinian freedom fighters struggled valiantly against their powerful Israeli foes.

The demonization process started when the Palestinians began to spread rumours that Israeli solders had in the course of battle deliberately slaughtered hundreds of innocent Palestinian civilians. They were assisted by the fact that the battle for Jenin had left a section of the refugee camp totally devastated.[30] And so, once the fighting died down, Palestinian spokesmen were quick to claim that the IDF had not only demolished numerous buildings, but also callously left masses of bodies buried under the rubble.[31]

The IDF's ill-judged decision to impose a news blackout on the operation and to bar reporters from entering the camp meant that little was known about what had truly transpired – a vacuum the Palestinians were quick to fill by adding to the rapidly growing spate of rumours and speculations, and supplying the media with interviews, pictures, and stories. Senior figures in the Palestinian Authority announced that some 3,000 Palestinians had been killed in the course of the fighting – a number they later reduced to five hundred.[32] In order to demonstrate that a horrific massacre had taken place in the town, The Palestinians exhumed bodies of people who died recently of natural causes and buried them in a makeshift cemetery.[33]

In order to arouse international interest in the story, the Palestinians took advantage of known sympathizers such as Terja Larssen, the UN special envoy to the Middle East. Dressed in a blue[34] protective vest and helmet, Larssen arrived on the scene shortly after the end of the fighting in order to investigate the Palestinian accusations. His quick arrival created the impression that there was indeed substance to the accusations.

A further sign of the Palestinian campaign's success was the decision of UN Secretary General Kofi Annan, on 19 April 2002, to appoint a special committee to investigate the events in the camp.[35] Israel claimed that it had attempted to reduce the number of civilian casualties: it had knowingly endangered its own soldiers' lives by choosing not to bombard the camp from the air, but instead to secure Jenin by close house-to-house combat; however, this claim received little notice in the international press, which was another measure of the Palestinian campaign's success. The massacre charges were finally laid to rest in August 2002, when Annan published a report stating that no such mass slaughter had taken place and that in the course of the fighting in the camp, fifty-two Palestinians had been killed, half of them civilians.[36]

## The Siege of the Church of the Nativity in Bethlehem

In April 2003, the IDF surrounded and set siege to the Church of the Nativity in Bethlehem, in which Palestinian militants had barricaded themselves. In the course of the siege, the Israeli army fired several shots into the building. The Palestinians exploited this fact and, appealing to Christian public opinion at large, launched a widescale demonization campaign in which Israel was accused of wilfully destroying both churches and mosques. The campaign lost something of its force when it emerged that the besieged militants had stolen invaluable holy artefacts.[37]

Presenting the world with a united front, the Palestinians disseminated simple, captivating messages, mainly that the crux of the Palestinian problem lay in the Israeli occupation, with the implication that its solution lay in ending that historical aberration. By exploiting most individuals' instinctive inclination to side with the weak against the strong, they emphasized the notion of a subjugated people fighting for its independence from a great regional power – an ever-popular theme in modern Western culture.

## PSYOP Delivery Channels

To the dismay of journalism theorists and practitioners, from the perspective of PSYOP, the media are a perfect channel to deliver messages to all designated target audiences, as the media are perceived in the West as relatively objective and fair. For the Palestinian activists, infiltrating the media proved to be relatively straightforward, due to the proliferation of private and public news and media agencies in the West Bank and Gaza. Developments in communications technology reduced the costs of producing and broadcasting stories, and this technology was quickly adopted by the Palestinian Authority and entrepreneurial individuals. Now the Israeli state was no longer able to censor content; the Palestinians filled the demand for information from the territories and made full use of opportunities to colour that information in accordance with their goals.

The shift from the Palestinians' alienation from the foreign media towards a policy of embracing it started in the mid-1970s. Arafat gave instructions to form links with numerous media representatives, offering them his full support and cooperation.[38] The Jerusalem Media and Communication Centre (JMCC), established in 1988 during the First Intifada and headed by Ghassan Hatib, is a perfect example of how the Palestinians built up a strong working relationship with the international press and used it for their own ends.[39] One of the activities of the JMCC was to contact journalists on its own initiative in their countries of origin and invite them on educational tours in Israel. It supplied journalists and filmmakers with background material and allowed them access to various "hot spots". It provided access to Palestinian and Israeli interviewees – only those with appropriate views, of course – and also furnished media representatives with much-needed guides and translators.

All of this, naturally, served to increase the odds that these overseas reporters would advance the Palestinian point of view.

In addition, during the course of the Intifada many foreign news desks, concerned for the safety of their reporters and beset by increasingly heavy insurance premiums, elected to remove their reporters and cameramen from the field and rely solely on Palestinian stringers instead. By choosing which stories to cover, from which angles to take particular photographs, and when to hand in their material to their employers, the Palestinian stringers became powerful foot soldiers in the Palestinian PSYOP campaign, ensuring that most of the stories published in the foreign press had a distinct pro-Palestinian twist.[40] News editors across the world were quick to embrace this new arrangement, which provided them with mainly visual material whose quality only improved as the Palestinian stringers honed their photography skills, though even jerky, amateurish video shots had their uses by creating the impression of real events recorded in real time: this injected a strong sense of drama into the news stories.

Generally, the Palestinians made sure that the reports that were sent out highlighted incidents that would advance their cause; and they were also careful to exclude reports about or photographs of any event that might prove detrimental to their cause, including the lynching of two Israeli soldiers in October 2000, and the exhilaration with which many Palestinians greeted the 9/11 attacks.[41]

## Propaganda by Deed

Unlike the First Intifada, which was essentially a non-violent campaign in which the Palestinians made little use of firearms, the Al-Aqsa Intifada, as mentioned above, was characterized by violence. The Palestinians deployed a range of weapons, including guns, Kassam rockets, road mines, and above all, suicide bombers. The result was heavy losses on both sides, with the Palestinians suffering some 3,000 dead, and the Israelis 1,100. The often spectacular Palestinian guerrilla and terrorist operations embodied the concept of "propaganda by deed" which had been formulated by the Russian revolutionaries of a century before, and became an important and integral part of the Palestinian PSYOP-related activities.

## Diplomacy

In order to influence world public opinion, the Palestinians established a vast, industrious, and professional diplomatic network. Inaugurated in the mid-1970s, it helped transform the PLO, long castigated as a pariah terrorist organization, into a legitimate political entity, whose goal of independent statehood enjoyed sweeping international support. Today the Palestinian Authority boasts 101 diplomatic missions, the vast majority of which are embassies.[42] It is an impressive number, which even a mid-sized country would be hard pressed to match. This massive diplomatic infrastructure was yet another tool utilized during the Intifada to promote the images and political campaigns of the Palestinians.[43]

## Web Sites

The internet offered the Palestinians ample means to disseminate their messages abroad, and they exploited this medium to the full. When searching refugee camps for armed terrorists, Israeli soldiers were often surprised when they stumbled upon an assortment of state-of-the-art computer equipment. With the financial help of Arab and Muslim organizations worldwide, the Palestinians launched a heavy on-line presence. Containing vast amounts of mostly visual information, a considerable number of these sites set about demonizing or blackening Israel's name in the eyes of world public opinion.[44] Palestinian universities proved particularly adept in utilizing cyberspace. One example was the University of Bir Zeit, which started early on to produce numerous web sites specifically directed towards neutral audiences.[45] It provided hundreds of links to other web sites that relate, directly or indirectly, to the Israeli–Palestinian conflict, such as web sites of journalists, filmmakers, and academics who had visited the university in the past.[46]

## Recruiting Activists Abroad

Through its support of such groups as the International Solidarity Movement, the Palestinians devoted much time and effort to locating and cultivating sympathetic young Westerners. Having been briefed on events from the Palestinian perspective, these young people were encouraged to come to the Gaza Strip and the West Bank, mostly as aid workers, or, in some cases, as human shields – with the unfortunate result that two of them became casualties of the conflict. In March 2003, an American, Rachel Corrie, was accidentally killed after tripping in front of an Israeli bulldozer as she attempted to stop it from demolishing a Palestinian house. A few months later, a British national, Tom Hurndall, was fatally wounded by an Israeli soldier when trying to lead Palestinian children from a firing zone to safety. Both incidents received extensive news coverage in both Britain and the United States and resulted in a barrage of anti-Israeli criticism.[47] The Palestinians made use of these sad events to further pillory Israel and its policies in the international arena.[48]

## Persuasion Techniques

In addition to traditional persuasion techniques, the Palestinians made use of two other techniques that deserve mention in this context.

## Manufacturing Incidents

The Jenin "massacre" was based on a real event, namely, the conquest of the refugee camp. But from the outset, the Palestinians utilized the technique of fabricating events – a well-established PSYOP tactic.

One such event was the death of the boy Muhammad al-Dura on the second day of the Intifada, at the Netzarim crossroads in the

Gaza Strip. Footage was produced showing him lying down after allegedly being hit by Israeli bullets as he and his father cowered behind an improvised shelter. While the image of his terrified face has become a symbol to the entire Muslim world, there is a growing body of evidence, albeit circumstantial, that published account of his death was a sham; that virtually the entire incident was a fraud contrived by the Palestinians as part of their PSYOP offensive. For example, the picture of the boy laid out in Gaza hospital's pathology lab and identified as Muhammad al-Dura bears no resemblance to the pictures of Muhammad released by family. The boy's uncle, who was incidentally the photographer who had captured the incident on film for the French FRANCE 2 television station, categorically denied in the initial stages of the investigation ever claiming that his nephew had died.[49]

After a libel lawsuit in France, FRANCE 2 was forced to release raw footage (rushes) of the incident which show just how the Palestinians went about orchestrating the incident. In one of these tapes, Palestinian activists can be seen spraying the air with bullets and then signalling to ambulances – which were spotted waiting around the corner – to rush to the "scene". The ambulances, having loaded the "injured" into the vehicles, speed away with their sirens at full blast. These dramatic events were, of course, caught on film by Palestinian cameramen, who happened to be passing by. The raw footage also captured, albeit inadvertently, the local audience, who, scattered along the sidewalk, could be seen heartily applauding the performance.[50]

The bogus funerals of the victims of the alleged Jenin massacre were another Palestinian attempt at manufacturing events, but in this case the attempt failed, to the great embarrassment of the Palestinians.[51]

### Personal Experience

This is a PSYOP technique used to influence members of neutral audiences by affording them either "firsthand experience of events" or, when this proves impossible, creating simulations that closely replicate reality. The Palestinians' link with the aforementioned International Solidarity Movement is a typical example of the first process, while the mock security fences (mini Apartheid Walls) set up on US and EU university campuses are a prime example of the second.[52] These interactive installations were manned by

Palestinian activists dressed as Israeli soldiers, who prevented visitors at random from entering the campuses simply on the grounds that they were Palestinians; these simulations attracted considerable public attention.

Aid workers were, of course, by the very nature of their work, routinely exposed to the genuine daily sufferings of the Palestinian people. The Palestinian Authority duly exploited the situation through briefings on the Palestinian civilians' plight and so heightened the aid workers' pre-existing feelings of empathy. It was a strategy that faltered somewhat, however, following the kidnapping of several foreign aid workers and journalists by militant groups. These groups, operating on a different agenda, hoped to coerce the hostages' governments into changing their political stance. Although the Palestinian Authority was generally quick to secure the hostages' release in an effort to limit the damage, the harm was already done.

## Conclusions

The Al-Aqsa Intifada was a classic example of a limited conflict between a small force and a large regional power. Certainly one of the reasons that the Palestinians gave so much weight to PSYOP-related activities within their overall strategy was that they had very little choice in the matter, given Israel's overwhelming military superiority. However, by making astute use of a variety of PSYOP techniques, they proved just how powerful a weapon it could be: it is a weapon that, when shrewdly combined with conventional military activity, can produce impressive political results.

By defining specific PSYOP aims, manufacturing messages tailored to individual audiences, maintaining a tight grip on all outgoing information, exploiting public relations opportunities and utilizing a wide range of persuasion methods, the Palestinians were able, at least to some extent, to counterbalance Israel's military preponderance – albeit chiefly in the political and diplomatic arenas.

One reason for this success was the expert way in which the Palestinian Authority identified and addressed the American and Western European audiences. The Palestinians learned during the First Intifada that careful identification of the important audiences could pay handsome political dividends, and now they further exploited their PSYOP skills and organization. Their messages to each selected audience were both verbal and visual, figurative and concrete. In addition, the media (old and new), private individuals, international organizations, and public institutions, together with their own diplomatic system, were all employed to get the Palestinian message across. At the same time, the Palestinians were careful to fashion their messages' content to suit the general, if oft-changing, political climate, as well as the cultural, social, and ethical mores of their target audiences. This was why the 9/11 attack, President Bush's refusal to negotiate with Arafat, and the suicide bombings in Israel did not, in the long run at least, seriously hamper the Palestinian PSYOP campaign.

The global media's fascination with the Israeli–Palestinian conflict, which led to an influx of foreign news teams into the area, especially during periods of heightened tension, was grist to the Palestinian PSYOP mill. As many news agencies increasingly relied on Palestinian stringers when the situation in the territories became hazardous for foreign journalists,[53] the Palestinians could supply the foreign media with material which, as illustrated above, was often misleading, to say the least. Cultivating their relations with the foreign and Israeli press, the Palestinians constructed elaborate machinery for distributing information, providing the press with newsworthy stories in real time. Employing modern technology, they used digital cameras, internet sites, and e-mail to relay huge quantities of mainly visual information to both the global media and the public worldwide.

No less important was the Palestinian understanding that Israel was bound in the PSYOP realm to rules of fair play. For the IDF and government bodies such as the Foreign Ministry, credibility was important, and they generally refrained from tactics such as inventing events. As a revolutionary movement, the Palestinians had no such compunctions, and they spread wild accusations and half-truths as matter of course.

As part of their campaign, the Palestinians exploited the fallout from the various political decisions and military measures Israel had adopted in response to the Intifada, as well as its blatant PSYOP-related mistakes. They made good value of the profusion of IDF roadblocks, which painfully disrupted the daily lives of most Palestinian civilians, subjecting them to countless major and minor humiliations, and of the bans on all movement between Gaza and the West Bank following particularly vicious terrorist attacks. Then there was the Israeli government decision to demolish all Palestinian workshops used for manufacturing Kassam rockets, even those located in densely populated areas – a decision which resulted in 150 Palestinian deaths, including civilian ones.[54] The systematic destruction of these workshops, as well as the razing of agricultural plots that provided cover for Kassam rocket launch sites and Palestinian ambush units, aggravated the Palestinians' already unstable economic situation – another good source of PSYOP capital. More importantly, the Palestinians benefited from the fact that with PSYOP playing a marginal role in Israel's overall security policy, it was unable to launch an effective PSYOP counterattack, even when the Palestinians' accusations were baseless.

Until 2006 the Palestinians proved adept at concealing their nation's deep internal political divisions and bitter personal rivalries, with Palestinian spokespersons projecting an image of a grimly determined and united people, resolved to fight to the end for its national liberty. The Palestinian Authority, which maintained almost complete control over all outgoing information, was particularly deft at weeding out any facts at odds with this upbeat picture of Palestinian society. Rival opposition groups, such as the Hamas and Islamic Jihad, appreciated the need to present a united front, and therefore tended to cooperate with the Palestinian Authority, generally refraining from airing their differences in public. Indeed, this display of Palestinian unity was not all show, as the Palestinian Authority used its PSYOP skills on its home audience as well, and with much success. Among other home-targeted campaigns, the PA launched one damning Israel as the sole cause of all the Palestinians' problems – thus divesting itself of responsibility for the disastrous effect of the Intifada on the quality of life of the average Palestinian, especially in Gaza. It thus managed, for a very long time, to deflect any local criticism and unite the Palestinian people under its auspices.[55]

As mentioned, the Palestinians were able to win several notable PSYOP and political victories. By launching a formidable media offensive directed at neutral and Israeli audiences alike, they seized control of the conflict's military agenda and, no less important, of its political agenda, with the result that Israel, much to its dismay, found itself constantly on the defensive. By showing the sufferings of the Palestinians at the hands of the occupying Israeli forces, they convinced neutral audiences to support the goal of an independent Palestinian state.

As a final point, it must be noted that the Palestinian PSYOP campaign directed at the enemy audience was even more persuasive. Immense pressure was applied by the neutral audiences on Israeli leadership through NGOs and European governments regarding human rights and international law issues. This pressure was exacerbated by exploiting the deep-seated ideological divisions within Israeli society regarding the fate of the West Bank and Gaza, which received much exposure not only in the quality press but also in the popular daily newspapers. The result was unbearable psychological pressure to retreat from the Gaza Strip. By 2005, most Israelis regarded the continued presence there as a liability rather than an asset, and approved of their Prime Minister's decision to disengage from Gaza unilaterally. Nor did it end there, since Israel had to retake the Gaza Strip as a result of constant bombardment by mortar fire and Kassam rockets on its southern region in 2008. PSYOP strategy is still pervasive with the Hamas government in the Gaza Strip as a result of the organization's military disparity with the IDF. Despite these experiences, many Israelis nevertheless support a withdrawal from the West Bank, a phenomenon that can be explained by the success of the massive Palestinian PSYOP campaign.

# 3

## From Oslo to Jerusalem
Fifteen Years of Palestinian Psychological Warfare against Israel (1993–2008)

The Oslo Accords (1993) between Israel and the Palestine Liberation Organization were the most significant political accomplishment of the Palestinians since they began their active opposition to Israel in the 1960s; it was also an important step, in their view, on the way to establishing an independent state. In their struggle to achieve political goals, psychological warfare was the primary means.

The Palestinians' widespread use of psychological warfare against Israel – in the past and at present – has been a primary reason for their success, and they have been more successful than other terrorist groups in their use of information warfare aimed at enemy audiences. What is exceptional about the campaign was that the activities, especially the violent ones – from suicide bombings in city centers to the shooting of Kassam rockets and mortar shells – were all designed to serve the political goals established by the Palestinian leadership; in other words, psychological warfare was the top priority.

This chapter will survey the Palestinian psychological warfare campaign of the last fifteen years, from the Oslo Accords (1993) to the end of the Tahadiyeh (temporary truce) between Israel and Hamas in Gaza as a result of Operation Cast Lead (December 2008–January 2009). The first part will focus on the principles that guided the Palestinians in this campaign. The second part will describe the specific strategies that they found suitable for use against Israel; and the third part will present a number of the techniques used in the effective delivery of their messages to the Israeli public – messages which brought them impressive political gains during this period.

In the study of psychological warfare, three target audiences are usually defined: the home audience, the enemy audience, and the neutrals. The chapter will focus on the Palestinians' efforts vis-à-vis their enemy target audience, namely Israel. "Enemy" is a broad term, and in this case includes all the influential groups in Israeli society. These influential bodies can be divided into two sub-groups: first, the government and educational, cultural, academic, and public figures; and second the local media (the foreign, i.e. Western, media is also included here, though not directly, through the pressure that it exerts on those parts of the Israeli public that are exposed to it). The two sub-groups have a symbiotic relationship with each other, and together can influence many segments of Israeli society.

The PLO leadership has used psychological warfare from the beginning of the struggle in the 1970s, but their organized efforts in this arena received new emphasis starting with the First Intifada, in 1987.[1]

## A Basic Definition

We define psychological warfare as "the sum of essentially non-violent means used during a conflict, in order to change attitudes and behavior of designated target audiences in order to achieve political and military goals". The important component of this definition is "essentially non-violent means", as it serves to distinguish between psychological warfare (PSYOP) and the psychological effect of a terrorist attack or a military action. The efforts to convince can take place not only during an actual war, but also during any prolonged conflict between two political entities such as states and terrorist organizations. Psychological warfare, by definition, is non-violent, for it is conducted through information dissemination, but PSYOP campaigns are often aided by military actions during a conflict (such as launching a missile, bombing from the air, or a terrorist attack), which also have a psychological effect on the target audiences.

PSYOP is an ideal warfare technique for the weaker side in a conflict, as that side lacks the resources and the military ability to engage in a conventional war against its stronger enemy. Concurrently, the stronger side will usually tend to neglect the use of PSYOP, as strong conventional armies tend to opt for short wars relying on their conventional military advantage. They also tend to view asymmetric conflicts as symmetric ones, and leave PSYOP to the enemy, thereby giving their opponents a great advantage.

The Palestinians recognized the asymmetry between their power and the Israeli military forces, and used PSYOP as a tool to achieve their goals by integrating professional knowledge acquired in the Communist bloc, and by learning through trial and error, thus improving their PSYOP operations.[2]

In order to communicate with the target audiences, the PSYOP initiator uses all the delivery means available to him, from leaflets dropped from airplanes to graffiti; from leaders' public statements to marketing technologies such as internet sites, SMS messages, beepers, and social networks.[3] Viewed from that perspective, the local and international media are merely another tool in the PSYOP agent's toolbox.

When one defines PSYOP, one encounters difficulties, for in conflicts there are also non-aggressive attempts to psychologically influence the other side. In any negotiation, for instance, both sides manipulate each other, using psychological tools such as threatening to leave the negotiating table, making various statements relating to the opponent's character and negative intentions, etc. To confuse the issue further, at the beginning of World War II in Britain psychological warfare was called "political warfare".[4] What, then, is the dividing line between negotiations and PSYOP?

Answering that question requires examination of all of the different methods used in the conflict. If political negotiations are just one means among others which including violence, the goal being to vanquish/destroy the opponent, then these negotiations are viewed as one part of an overall PSYOP strategy.

In the Palestinian case, the goal of the Palestinian Authority was to force Israel, through mutually supportive military and political moves, to accept a Palestinian state alongside it; according to the statements of its various factions, this state is to be a forerunner of "Palestine from the [Jordan] River to the [Mediterranean] sea". That places the Palestinian Authority's actions within the realm of PSYOP.

From this viewpoint, the political negotiations supported the PSYOP but also vice versa, and so, in various places in this chapter, some elements from the realm of legitimate political negotiations will indeed be included as parts of PSYOP campaigns and will be analyzed as such.

The Palestinians use PSYOP, then, on two levels in this conflict: strategic and tactical. On the strategic level, it is a central aspect of their political activities and its aim is to facilitate the achievement of political goals. The use of violence is subservient to their PSYOP strategy, and indeed is an integral part of it. On the tactical level, PSYOP is as important as all the other branches of activity. It intensifies the effect of the military operations, and together they serve Palestinian political goals.

## Operational Arm

In a famous speech in 1996, Arafat stated that he would destroy the Zionist state through an Arab population explosion and psychological warfare.[5] Even if this did not constitute official Palestinian policy, the fact that Arafat mentioned the topic indicates his awareness of the importance of PSYOP on the strategic level in the Palestinian struggle against Israel.

There is very little information available about the Palestinian PSYOP organization. It is known that there was such a body in Israel as part of the Intelligence Department of the Israel Defense Forces, namely the Intelligence Warfare Branch, which was disbanded in 1999 and reassembled in 2004, and is now called *Malat* (an acronym for *Mercaz Lemivtze'ei Toda'ah*, or Center for Consciousness Operations).[6] But the Palestinians did not release any information about their PSYOP organization until Operation Cast Lead in December 2008, at which point Hamas reported that it was attempting to neutralize Israeli PSYOP using various means such as hijacking the Israel Defense Forces Radio Station Galei Tzahal in the southern part of the country, putting out leaflets in Hebrew, and operating a system of websites for various audiences.[7]

As in all Arab states, the Palestinian Authority appointed a Minister of Information – Saib Arikat, who rose to eminence in the First Intifada – immediately upon its establishment. Arikat had accumulated much experience and had served in various spokesperson positions for many years. One cannot ignore the difference between this and Israel's approach, which has largely been to ignore opportunities for PSYOP: the Israeli government had a short-lived Hasbara Ministry in 1974, and now, almost four decades later, the Netanyahu government has created a Ministry of Information and Diaspora Affairs, but it has little real power.

At the beginning of the Second Intifada, the PA announced the opening of courses in PSYOP, the goal being to undermine the alleged Israeli superiority in that realm; in fact the allegation that Israel was effective at PSYOP was itself a PSYOP maneuver. In his "victory speech" at the end Operation Cast Lead, Ismail Haniyeh complimented the Gaza residents for "having bravely withstood the Israeli air-force, two divisions of infantry and hundreds of soldiers in the Israeli psychological warfare unit". The *Malat* people must have suppressed a sigh at this overestimation of their manpower. Until recently, Hamas was in the early stages of its PSYOP development, but the meticulous organization of the flotilla from Turkey in 2010 shows that it understands the importance of influencing the neutrals.

## Delivery Channels

A PSYOP campaign requires various means of delivery in order to pass information to the designated target audience. The choice of means is dependent on their availability and resources, and they include a wide range of possibilities, from rumors to airborne television stations.

The Palestinians have always shown remarkable creativity in choosing and creating ways to disseminate messages, using the technologies available to them and overcoming the obstacles Israel has presented them. In the First Intifada they used leaflets published by the Unified Leadership and fax messages, and they also coordinated to broadcast the same messages on the loudspeakers of mosques, transforming them into something of a makeshift radio station. Home video cameras, which had just entered the market, changed the way conflicts were covered, and the Palestinians (with encouragement from the foreign media) were quick to adopt this new technology.

At the Oslo Accords, Arafat demanded a broadcasting station, and since then Palestinian TV and radio stations have become an especially effective tool for the dissemination of messages to the Palestinian home audience.

Since the middle of the 1990s the internet has become an extremely important tool for disseminating information. It crosses borders, is inexpensive, and is very effective in disseminating messages to segmented target audiences.

Several factors contributed to the increased use of the internet as a tool in the Palestinian PSYOP campaign: the Palestinians are highly educated, there is a Palestinian diaspora in the Western world, internet technology is readily available due to the proximity to Israel, and content on the internet is not censored. The Palestinians are massively represented on the internet. The Palestinian universities such as Bir Zeit and Al Najah are "website-producing factories".[8] This is a coordinated Palestinian effort,[9] and not a spontaneous movement, and therefore fits within the definition of PSYOP.

The web enables the dissemination of incendiary information, such as the claim that Israel is using Palestinian prisoners as guinea pigs for the testing of chemical and biological weapons[10] (see the section titled "Demonization" below). A further advantage of disseminating information through the internet is the speed and wideness of its diffusion, and thus its power: in the case just mentioned, a short email resulted in a general alert among Israeli prison authorities and caused the Ministry of Foreign Affairs' spokesman and the Civil Command to address the issue (see below, "Overburdening the emergency and security systems").

The anonymity provided by the web enables a broad and rich range of black or gray PSYOP – which in the past had demanded more resources and sophistication. For example, the Egyptian media reported in 2008 that Gilad Shalit had been injured in one of the IDF's bombings of the Gaza Strip in Operation Cast Lead, basing its report on a quote from an Islamic website.[11] This false rumor received widespread coverage during the Operation.

## The Overall Goals of the Palestinian Campaign vis-à-vis Israel (1993–2008)

The Palestinians ascribe great importance to all three PSYOP target audiences: enemy, neutral, and domestic. Israeli society, as the only Western democratic society in the area, is a society that goes to the polls, influences its representatives on a daily basis and provides the manpower for the security forces. The neutral audience, especially the US and Western Europe, can apply political and economic pressure on Israel in order to force Israel to accede to the Palestinians' political demands. As mentioned above, this chapter will not directly address the neutral audience, except in the context of its applying pressure on the Israelis, who are concerned about their image in the West.

Beside civilians, the home audience includes Palestinian activists and soldiers (and their commanders). The importance of the home audience in PSYOP doctrine lies in the understanding that a basic condition for success is the population's willingness to join or support the struggle and duly sacrifice whatever it takes to achieve victory. These things are particularly relevant for the Palestinians, since their conflict with Israel is difficult and prolonged, and they have suffered much damage and loss due to Israel's superior military capabilities.

The overall goals of the Palestinians vis-à-vis the various audiences were determined by the Palestinian leadership, headed by Yasser Arafat – leader of the PLO and later Chairman of the Palestinian Authority for a period totaling four decades, until his death in 2004. Until Operation Cast Lead, the Hamas leadership, headed by Ismail Haniyeh, which has ruled the Gaza Strip since June 2007, viewed armed conflict against Israel (by way of suicide bombings and Kassam rockets fired at Israel) as the main tool for achieving its goals. Since January 2008, the leadership has invested effort in stockpiling armaments in the tunnels in Rafah area,[12] and at the same time attributes much importance to PSYOP, the Goldstone Report being one of its main achievements.

This section will deal with the Palestinians' messages to the enemy audience – Israel, especially Israeli civilians. A survey of the Palestinian PSYOP messages shows that the vast majority were intended for Israeli civilians, and fewer to the army, since any direct appeal to Israeli soldiers would certainly be jeered at and rejected. The obvious solution was to target the policy-makers and public-opinion formers, who would in turn create new guidelines that would limit the army's movements.

The main political goal of Palestinian PSYOP was to convince the Israeli public and its leaders – directly and indirectly, through neutral target audiences as well – of the need to implement the Oslo Accords of 1993 and 1995 and the Wye Agreements of 1998. These were meant to bring about a gradual withdrawal of Israel from Judea, Samaria, and the Gaza Strip, and to expand the Palestinians' rights accordingly, with the process ending with bilateral negotiations on a permanent settlement of the conflict. In addition, the Palestinians wanted to garner wide political support, both in Israel and abroad, for their policy, and to convince the Israelis of their sincere intention to uphold the agreements, to be moderate, and to stop the terror attacks. They also were working towards the release of the prisoners held in Israeli jails (11,000), while maintaining economic ties with Israel and receiving various public services including as medical service, technical assistance, etc.

The goals of the Palestinian PSYOP vis-à-vis the Israeli military were not different from those of other national struggles: to weaken their morale (by claiming that the Palestinians' determination and willingness to suffer would ultimately bring about their triumph); to encourage guilt in Israel over the harm caused to Palestinian civilians; to deter Israel from engaging in military actions against the PA and other factions in its territory in response to terror attacks, especially during the Second (Al-Aqsa) Intifada of 2000–2005. In order to accomplish all this, the Palestinians combined non-violent persuasion methods with the psychological impact of violent terror attacks.

## Principles of Operation

Theoretically, after drawing up a list of political goals, the PSYOP agent chooses from a wide range of tactics that can suit his needs. Yet it seems that Arafat never had a carefully thought-out plan, but rather proceeded based on trial-and-error that developed over several years. Research on the subject shows that the Palestinians did not invent a unique operative doctrine, but rather adapted and updated existing ones and put together a suitable, specialized method tailored to their specific conditions. They had many sources to work with, namely all the techniques used in the wars of the twentieth century in struggles against democracies, such as the pre-independence struggle of the Jews against the British Mandate, the Algerians' fight against French rule, and the conflict in Northern Ireland.[13] But the Palestinians did deploy one new strategy, which is that of suicide bombers.

The main operational techniques of the PSYOP used by the Palestinians against Israel included guilt, ostracizing, linkage, conditioning, deterring the enemy through cruelty, using concessions as a springboard, convincing the enemy that assets are liabilities, inflated power, and the new leaf syndrome. These techniques are explained in the following pages.

## Guilt

Guilt is one of the major motivators of human behavior, and PSYOP makes good use of it. Instilling guilt in the enemy soldier during war severely degrades his efficiency, and when guilt is instilled in the civilian population that sent him, the soldier is indirectly influenced as well. Moreover, guilt can affect the entire worldview of the young men and women who are going to be drafted in the coming years. Guilty feelings arouse thoughts of pacifism and unwillingness to serve in the army for reasons of conscience. In the meantime, the PSYOP operator uses every opportunity to arouse motivation for continued fighting in his own home audience.

The First Intifada was a turning point in the Palestinian PSYOP strategy, for it was then that the Palestinians identified the Israeli public as a target audience that could produce political achievements for the Palestinians. Various techniques were used in order to show the Israeli public the ugly face of the war – especially civilian casualties. The Palestinians disseminated stories and pictures of civilian casualties; one example is the civilians injured and killed in the course of the bombing of Salah Shehade's house in Gaza in 2002.[14] An important target audience for these PSYOP messages was the Israeli Air Force pilots, as well as the artillery corps, both of which operate from afar (stand off weapons) and prefer not to see the results of their actions.

## Ostracism

The goal of this technique is to damage the relationship between the enemy's government and the international community. The PSYOP agent uses both "soft" connections (culture, media, social) as well as "hard" (diplomacy, security, finances) to achieve this goal. Once the image of the enemy is harmed, the operator will try to extend the damage in other realms. Much energy was directed since the 1970s to pass anti-Israel resolutions in international forums. A striking example was the 1975 UN 1975 "Zionism is Racism" resolution,[15] and more recently Israel was accused of committing holocausts and being anti-Semitic at the first Durban Conference (2001). Israel is constantly being embarrassed in diplomatic gatherings; symbolic economic boycotts are declared, such as on the Dead Sea Ahava products, and economic tools are used to sabotage Israel's security, such as in the campaign against Caterpillar Inc., the manufacturer of bulldozers.[16]

The ostracism effect can snowball, since after a number of early successes, the target state's image becomes tarnished, and new accusations are more readily believed, making it easier to achieve political goals against the enemy.

### Linkage

The principle involved here is the claim of a causal connection between the actions of the two sides, in a way that is favorable for the initiator of the most recent action. For instance, "We acted so because of what you did". This technique was used mainly to influence the decision-makers in Israel, as well as the general public; its intention was to create discord in Israeli society, as some people will adopt the Palestinian narrative and claim that if the Israeli leaders had not acted as they had, the Palestinians would not have responded as they did. Another aim is to limit the leaders' room to maneuver by placing a check on their initiatives or responses, namely, the possible reaction of the enemy. At first this way of thinking takes the form of apparently rational reactions to specific events ("This terror attack could have been prevented if only..."), but as time goes on the idea sinks into the subconscious, becoming a paralyzing influence that increases feelings of helplessness and defeat. On a secondary level, this principle can be used for countering criticism from the Palestinian home audience: if Palestinians were injured during an Israeli attack that came as a result of a Palestinian attack, the Palestinian leadership could justify their actions by saying they acted in response to an Israeli move.

One notable implementation of this principle was Hamas's announcement that it would execute revenge terror attacks in response to the assassination of Yahyah Ayash, known as "The Engineer" of Hamas, in January 1996. In a report to Reuters, Hamas specified that ten suicide bombers had been trained and were awaiting orders. The announcement was indeed followed by a double suicide attack in Beit Lid.

The reality was that there was no connection between the two events, for an attack of that scope requires lengthy planning and preparation. The main goal in this case was to deter the decision-makers in Israel from acting against Palestinians in the future, with the implication that there would be a similar response to any act they took. ("If you had not acted, the response in the form of a terror attack would not have happened either"). Even afterwards, Hamas did not stop issuing warnings about attacks. The fact that few if any of them never came about did not stop the Israeli public from believing Hamas (especially when the Israeli security forces supplied some confirmation of the danger). Thus, the Palestinians succeeded in instilling in the Israeli public a belief in their dedication to revenge, and when they vowed vengeance after an Israeli attack, the Israelis accepted and believed that that was the way of the Arab world, and that a string of terror attacks was inevitable.

### Deterrence through Cruelty and Determination

The militarily weaker side in a conflict does everything it can to prevent its powerful opponent from activating its superior military force against it, for it is clear that in a conventional military confrontation the weaker side would be defeated, and its political achievements would be undone as well. An old technique for achieving this goal is to deter the enemy by creating a façade of determination strengthened by harsh measures. The determination is demonstrated by showing a widespread willingness for self-sacrifice in the national struggle: interviewees talk about how many relatives they have lost in the course of the armed struggle; old people and children make statements about being willing to fight to the end.

In the case of the Palestinians, these interviews generally took place in refugee camps, in order to emphasize their dire poverty – implying they have nothing to lose. The theme of "Palestinian determination" has been further played up by leaked stories about how Hamas trained suicide bombers by burying them alive to test their resolve. These reports received coverage from several media outlets in Israel and in the US.[17]

This particular doctrine is part of a paradox that permeates all Palestinian strategy. On one hand the Palestinians wish to present themselves to the Western world as David battling Goliath (hence the great efforts in branding the conflict as "Israeli–Palestinian" and not "Arab-Israeli"); on the other hand, they wish the Israelis to fear their military strength. In fact Israel faces the same challenge, trying to show its weakness to the West and its strength to the Arab world. Yet, the Palestinians have managed to simultaneously project these two contradictory images much more successfully.

The answer to the weakness/strength paradox is multi-level. On the psychological level, people can believe contradictory things. On the ideological level, in the West in the 1950s the perception developed that when it comes to wars of independence in the Third World, all ruling powers were bad and all national independence movements were intrinsically legitimate. The Palestinians filled in the missing piece of logic, namely, that Israel is an offshoot of a colonizing world power, i.e. America, and that therefore Palestinian struggle is legitimate, whereas the Jewish struggle was colonial and reactionary.

Once the Soviet bloc collapsed, the Palestinians focused on the ideological foundations developed during three decades of radical discourse on human rights. When Hamas took over in Gaza, it too continued this split discourse, creating media events such as fake blackouts in Gaza,[18] and exaggerating the number of civilian casualties in Operation Cast Lead[19], while continuing its Islam-based demonization of Israel, claiming that Jews are the descendants of apes and pigs.

Another split that the Palestinian leadership, specifically Arafat, succeeded in creating was that between the suffering Palestinian people and small violent groups within it that have no choice but to act.[20]

The cruelty came to the fore during the Second Intifada, following Israel's successful targeted assassinations. The Palestinians removed the bodies from the burnt vehicles and presented to the media pictures of inner organs; they dipped their hands in the blood of the dead and held up clenched fists. This activity conveyed the message that "We're lunatics when it comes to our national struggle." It was a very powerful visual image, and the pictures appeared both in the Israeli and the foreign media. But soon afterwards the Palestinians stopped showing such pictures, having understood that they actually served Israel's interest by showing the Palestinians as a savage, uncultured people. Instead, pictures were taken of Palestinians in demonstrations carrying swords and wearing fearful masks – thus conveying the message that "We are savages and our vengeance is terrible".[21]

Another expression of this principle was the merciless treatment that was given to those who were accused of collaboration with the Israeli enemy when Hamas was coming to power in Gaza (2007). After a hasty trial came the executions, which were held publicly and covered by Hamas media. The videos of the executions are very violent, and therefore were not publicized in the elite media such as international networks. They were intended for Israel, and also for the Arab world.[22] The message expressed by such acts was that the Palestinians are merciless, desperate, and totally uninhibited when it comes to achieving their goals – and therefore unstoppable. These are very effective messages that serve the purpose of unnerving the opponent and making him want to "have as little contact as possible with these lunatics" – a wish that will be translated into political concessions.

## *From Asset to Liability*

It is a central principle of PSYOP to convince the enemy that the perpetuation of his present policies will cause more harm than good. Harm could mean further terror attacks, low morale, damage to his international connections and image and, economic implications. This principle was used mainly with the Israeli target audience, but also with the neutrals (the US and Western Europe). Israel received messages regarding the financial cost of the occupation, the moral price, and the damage to Israel's relationship with other nations.

The Palestinians addressed audiences in the US, including Jews,[23] with messages that tarnished Israel's image; they then turned to the Israelis and showed them their image abroad. This method produces great psychological pressure, but also leaves a way out: if Israel just changes its policies, it will redeem itself politically and the international pressure ought to decrease accordingly.

## *Inflated Strength*

This is a well-known principle in PSYOP operations, and is similar to the principle of deception, in which one of the sides exaggerates its own power. The goal of this is to astonish the enemy, make him despair, and force him to change his policy or to refrain from engaging in some planned military action for fear of heavy losses or of the inability to achieve a quick victory. All this can cause increased use of resources in the long-term and decreased decisiveness.

The practical applications of this principle by the Palestinians were the presentation of exaggerated data regarding the size of their population (see the subsection below titled "Surveys"), the numbers of fighters in the various militant organizations, the quantities of weapons and equipment they possessed, and the strength of fortifications and obstacles.[24] The show of inflated strength may be done by way of public announcements, but occasionally in more roundabout ways (holding large rallies, showing videos of training camps, etc.). This was how the Palestinians cultivated in the IDF the fear of a Palestinian military action during the Second Intifada while hundreds of Israelis were killed in suicide bombings.[25]

A similar phenomenon was seen in December 2008 at the end of the Tahadiyeh, when Hamas publicized information about the fortifications it had built and the arms it had smuggled into the Gaza Strip in an attempt to deter Israel from attacking it.[26]

### New Leaf Syndrome

It is a long tradition in the Middle East – and an old element of Islamic war strategy – that when one of the sides in a war, usually the weaker one, wishes to end a violent chapter in a longstanding conflict, it can announce its wish for a "clean slate". In this "new page in our relations", the past violent actions are (apparently) to be forgotten.

Such a hiatus is crucial for the losing side, for it gives it an opportunity to recover and reorganize, and then renew its efforts (in the form of more violence). This technique is apparently based on a similar principle that was used by Arab tribes, and was intended to give both sides some periods of quiet in unending conflicts. Announcing a "clean slate" is meant to convince the enemy to lower its psychological defenses and enable the announcer to mobilize people and organize resources.

The extent of Israeli willingness to go along with these moves was demonstrated during one such period in 1993. The IDF opened a special school for cooperation with the Palestinians, in which research was done on negative stereotypes, and classes were given on such matters as bridging cultural gaps, etc., and other matters crucial for the running of the joint patrols and Israeli–Palestinian Authority liaison headquarters.

During each "clean slate" initiated by the Palestinians and the quiet period that followed, Israel agreed to warm up the relations with the Palestinians. The Gaza checkpoints were opened and a number of checkpoints in the West Bank were removed; trade was renewed; joint dinners with liaison officers were held; and political contacts were made.

In one such period, Abul Abbas, member of the Popular Front for the Liberation of Palestine of Ahmed Jibril, returned to Gaza and expressed his sorrow (April 1996) over the attack on the *Achille*

*Lauro* cruise ship (October 1985), in which the elderly Leon Klinghoffer had been thrown overboard – for a new era had now begun.[27] All in all, the Israeli public was thrown into an atmosphere of total political novelty where old truths were cast aside. The Palestinians were quick to push for more concessions from the Israeli government and where the suicide bomber campaign immediately followed it was too late for the government to retract. The blame was put on the radicals (Hamas and Islamic Jihad) as intransigent elements, whereas the newly erected Palestinian Authority[28] headed by Arafat was considered a legitimate partner.

## Palestinian PSYOP Strategies

In order to convince its target audience – the Israeli public – the Palestinians consistently used several strategies that guided the message delivery and any other activity meant to further the achievement of their political goals. These strategies included initiative and attack, gradual increase in demands, selective use of the truth, developing a relationship with the media, and the use of public relations techniques. They were thus able to get the most out of every incident and exploit Israeli feelings.

### 1. Initiative and Attack

In PSYOP, as in conventional warfare, initiative is the key to success, the goal being to keep the enemy in a defensive posture in the face of an unceasing "bombardment" of messages, in times of quiet as well as in times of tension and open conflict; the enemy must not be allowed to relax for a minute.

On the strategic level, the constant attack is intended to convince the opponent that any protest on his part about these activities is futile and will only bring about embarrassment and damage, therefore if he wishes the incessant pressure to stop, he should stop resisting and abandon all efforts to win – first in the realm of information dissemination, and afterwards in the military arena as well. On the tactical level, the constant attacks force the enemy to apologize and justify himself constantly, leaving him no time to plan his own attacks properly.

The implementation of this principle requires daring, much emotional energy, and the willingness to pay the price of criticism in case the attack fails. The Palestinians used this principle even though they were perceived at first as idiots by Western audiences, and they certainly had no qualms about using it vis-à-vis the Arab audience and Third World countries. A case in point is the behavior of Iraq's Minister of Information, Muhammad Saeed el-Sahhaf, who during the 2003 invasion of Iraq repeatedly made grandiose false claims about Iraqi successes on the battlefield, constantly remaining on the propaganda attack, and as a result mistakenly labeled by the West as a clown on the verge of insanity. Closer to home, after the interception of the *Karin A* in 2002, Arafat denied any knowledge of a weapon-carrying ship. When the head of the IDF Information Branch said (January 1995) that a process of "Lebanonization" was beginning in the territories, Arafat responded immediately by saying that the actions of the head of the Information Branch were the cause of it: he was always ready to attack with a new claim.[29] Another example is the theme that the Palestinians are the direct descendents of the ancient Phoenicians, which was designed to undermine the Jewish claim to Israel. Indeed, there is ample room for an un-politically-correct, in-depth, cultural-anthropological study of Arab rhetoric, views of reality, and shaping of realities.

Determination and repetition contribute much to the success of PSYOP actions, and even messages that seem completely ridiculous at first glance become, after a long-term campaign, popular and accepted. See below for some examples.

## 2. Interfering with Israeli Activities Abroad

The Palestinians wish to prevent Israel from undertaking any activities abroad whatsoever, even though this might seem a violation of the accepted rules of international behavior. Over time, this campaign scored significant successes. For example, the organizers of the annual international book fair in Turin found themselves in a difficult situation following their announcement (March, 2008) that Israel was to be the guest of honor in celebration of the sixtieth anniversary of its independence. A torrent of protest arose among Palestinians, including demonstrations by Muslim writers' organizations and various fronts such as the Camden (London) Abu Dis Friendship Association, and J-BIG (Jews for Boycotting Israeli Goods). Demonstrations and protests took place opposite the Italian Embassy in London and in various places in Italy.[30] Over the years, the Palestinians have thus created reluctance in Europe and in the US to involve Israel in internation events because of the consequent controversies and embarrassments, and as a result, many activities supporting Israel have been relegated to Jewish communities and hard-core supporters.

## 3. Demonization

The atrocity propaganda against Israel is intended to present it as a satanic state that performs criminal acts against the Palestinian civilian population in violation of the norms accepted among other nations. When Hilary Clinton visited the Gaza Strip in November 1999, Suha Arafat told her that Israel was poisoning the Palestinians with gas.[31] In another case, the Hamas Ministry of Health in Gaza announced (January 2008) that it had to decide between supplying electricity to maternity wards and heart surgery operating rooms because of the lack of fuel caused by the Israeli economic siege.[32] In another case, they claimed that Israel's shutting off of the fuel supply was preventing the operation of the incubators in the neonatal units.[33] In spite of the fact that any thinking person would realize that Hamas needed only to refrain from launching rockets at Israel in order to receive everything its people required, the Palestinians wished to create a general impression of Israel as a cruel, evil state that causes much suffering to the local civilian population.

## 4. Maximizing Effects

In order to maximize the psychological and political capital from every incident, the Palestinians made sure to keep each incident in the spotlight for as long as possible. Any event, such as a demonstration, procession, military action (targeted killing), or civilian move (a closure), was presented from a variety of angles. In the case of a Palestinian death, interviews were held with the deceased's mother, brothers and sisters, friends, and commanders in the organization (occasionally a secret interview); artifacts were brought out to prove to the reporters that he was a normal, innocent person. If an entire family or group was involved, this process was multiplied according to the number of people.

In order to get maximum political gain from the victims in the national struggle, whether they were suicide bombers or civilians killed by mistake by Israel, a ritual was set for the funerals, which were Islamic in character. The body was covered with a green cloth but the deceased's face was left exposed, and there was a ceremony of leave-taking with his family. When several people had been killed, there was a joint funeral; the bodies were presented to the media for a photo-op, and the funeral procession was given full media coverage. The authentic rage accompanying the funeral produced strong images for the Palestinian photographers who since the year 2000 (the beginning of the Al-Aqsa Intifada) have taken over coverage of the territories, especially the Gaza Strip.

## 5. Gradual Increase of Measures ("yanking the chain")

One technique of revolutionary warfare is based on taking small, imperceptible, non-threatening steps over a long period of time. Every step is based on the previous achievements and itself serves as a springboard for another step forward. When the enemy eventually realizes the situation it may be too late already, both politically and militarily. This strategy is intended to prevent the enemy from reacting strongly and decisively. As a result, if on occasion the pressure on the enemy becomes too great, one must decrease it, otherwise a painful reaction by the enemy might result, upsetting the balance and eliminating all the advantages achieved thus far.[34]

Here one must differentiate between legitimate political activity and psychological warfare. The principle of gradual increase of measures is also used in routine political activity and in any negotiation, but – as mentioned above – when this gradual increase is supported by violence or integrated with it, and is only one means of many used to bring about the complete destruction of the opponent, rather than just political gain, then such activity should be considered as part of the realm of warfare.

The Palestinians used this strategy consistently, mainly in order to erode Israel's rule in the territories, and in order to gradually abet the achievement of their overall goal of an independent state. They went about this task as one would any complex project: by breaking it up into sub-tasks down to the smallest details.

When the PA leadership felt that the suicide bombings were "yanking the chain" too much, and that Israel was on the verge of reacting strongly, they took several measures in order to lessen the tension. General Razi el-Jibli, commander-in-chief of the Palestinian Police, was photographed with a potential suicide bomber that the PA had caught before he had left for his suicide attack on the settlement of Kfar Darom in the Gaza Strip. Arafat announced (June 1995), that the PA had captured thirty kilos of explosives intended for a terror attack.[35] Hamas announced (June 1995) that in the attack in Gush Katif, the terrorist had acted on his own initiative[36] (see also the section "Evasion of responsibility" below). At the same time, the Palestinians were attempting to deter Israel from taking any measures (such as targeted assassinations carried out from the air, sending ground troops into the Gaza Strip, or closing the territories) by voicing threats about the renewal of suicide bombings, emphasizing the large number of Israeli casualties that entering Gaza Strip would cause, and noting that closing the territories is not effective as the Palestinians have ways to get around it.[37]

From the moment the Palestinians achieve a victory in negotiation, that result immediately becomes a given, and the starting point for the next demand.[38] Although this is a principle used in many negotiations, the Palestinians took this technique one step further, creating the feeling that whatever was decided had always been the accepted view and that there was no room whatsoever for rethinking. In such a situation, there are no red lines to the negotiation, and the goal is to vanquish the opponent – a zero-sum game.

One Palestinian action in keeping with this principle was their demand that Israel release prisoners as a show of good will. Then, when 101 prisoners were released following the Oslo Accords, the PA limited the scope of the celebrations, stating that too few prisoners had been released, and demanding that more be let out, including those directly involved in fatal terror attacks.[39] (Israel holds some 11,000 Palestinian prisoners.) It also demanded that Hamas leader Sheikh Ahmed Yassin be released, but Israel refused in this case to comply.[40]

The Palestinians disseminated rumors regarding Israel's future moves: "Israel has promised to release 3,500 prisoners", "Israel will have to agree to the return of 100,000 1948 refugees to the Galilee".[41] Such activities go beyond the narrow spectrum of psychological techniques used in negotiations, and are part of a general strategy of tiring out the opponent, and of instilling despair in the face of a cleverer and more skilled opponent, as well as the feeling that time is on the opponent's side.

## 6. Attitudes towards the Truth

In Western Judeo-Christian culture lies are considered negatively and should be avoided as much as possible. When a leader is caught lying, that in itself is reason to call for his or her resignation. On the other hand, in the Arab world (and the Far East) maintaining one's personal honor (face) is much more important, and justifies far-reaching steps, including the use of varying degrees of truth.[42]

Thus no one in Palestinian society was the least bit embarrassed when it was discovered that a statement was not consistent with reality. The Palestinians under Arafat did understand from the beginning of their political campaign that it was not advisable to be caught in a lie when the Western countries were concerned. Hence the doubletalk strategy which was developed by the PLO since the late 1960s: a moderate statement directed towards the West, and quite a different statement directed towards home or Arab audiences. The Palestinians developed considerable skill in manipulating statements so as not to offend either audience. For example: "The Palestinian casualties of Israel's attacks are innocent, and any harm that befalls the Palestinians is the result of a deliberate action on the part of Israel". This claim does much to limit the scope of Israeli military activity, to increase its guilt for harming innocents (see below), to blame any escalation on it, and to present it as committing war crimes (see demonization and attacks).

When the targeted assassinations of senior terror perpetrators and Kassam rocket launchers in the Gaza Strip began, the Palestinians claimed that Israel was harming innocents.[43] In such reports, the main emphasis was on the civilian casualties – the passers-by and especially the children who were killed – but the launchers' practice of using civilians as human shields when driving and during military actions went unmentioned.[44] Naturally, the fact that the assassinated person was a member of a terrorist organization was denied (unless it was a known military leader).

If there was no choice, the victim was presented as a member of a terrorist organization, but one who was active in the political or religious wing. When the Palestinians provoked Israel to attack, for example by launching rockets from a residential area, the Palestinians neglected to mention that aspect of the story and emphasized the civilians harmed. Since in many cases armed insurgents acted from civilian-populated areas, the intention being to enlarge the circle of victims (as part of the message to their home audience), and many of the Palestinian fighters did not wear uniforms during their activities, it was easy to present all of the casualties, including the insurgents, as civilians. Presenting matters this way had a cumulative effect on audiences both in Israel and in the West.[45]

*From a lie to conventional wisdom*  The Palestinians adopted a few tactics that are reminiscent of Goebbels' "Big Lie" technique. Repeating messages – as obviously ridiculous as they may be – eventually turns them into conventional wisdom, especially if the message is strengthened by supporting bits of information. In this way, Israel and the Occupation could be blamed for anything that befalls the Palestinians, and peripheral topics can become major issues in the political discourse. The most impressive Palestinian achievement in this arena is the dissemination of the notion that "Israel is somehow responsible for this big mess" which is prevalent in global political discourse.

*Evasion of responsibility*  Any time an event bore the potential of harming Palestinian interests, steps were taken to evade responsibility and to blame someone else, usually Israel. Such claims have caused Israel to despair, as one cannot conduct a rational discourse with an irrational party. There are many advantages to this method, and it is dependent on the perpetrator's willingness to be perceived (at least in the beginning) as ridiculous. The use of this method increased during the Second Intifada.

While evasion of responsibility is a common political technique, the Palestinians refined it for the realm of PSYOP, making it into an operational doctrine. In a number of cases of Palestinian violence, they found an angle to the story that enabled them to minimize the damage to their image, and even to turn it around and use it against Israel. In the case mentioned above, it was the driver who drove full speed into a crowd waiting in a bus stop at the French Hill junction in Jerusalem, and was shot to death, but the Palestinians claimed that the Israelis were thoughtlessly confusing a traffic accident with a terror attack.

This is an example of working towards the strategic goal of the violent Palestinian campaign: to make life unbearable for the Israelis. Another point evident in this incident is that the target and details of the operation (choosing an Arab-American driver) indicate that the incident was well planned PSYOP-wise. The fact that the perpetrator was American could add much to the story's resonance in the US ("American citizen shot in Jerusalem by an Israeli soldier", etc.) The more common use of this method was blaming a violent terror act on a front organization or on a "loose cannon" – an individual who disobeyed the organization's orders, yet his/her act was a result of Israeli atrocities, etc.

***Arafat's lies*** After the Oslo Accords, Arafat, as the head of the PA, attacked Israel ceaselessly, accusing it of a multitude of crimes. When he was confronted with the truth, he denied everything and immediately went back to attacking Israel.

A number of examples: when Arafat was asked to respond to the attack in Beit Lid (March 1995), he said that the terrorists went through five or six Israeli roadblocks and no one stopped them, therefore it was Israel's fault.[46] In May 1995 Arafat presented 1,500 blank Israeli ID cards captured in the Gaza Strip, which, according to him, had been sent by Israel to criminal and terrorist figures in the Strip.[47] He claimed (March 1996) that the devastating attack on the no. 18 Jerusalem city bus was perpetrated by Israeli minister Rechavam Ze'evi.[48] When the *Karin A* armament ship, bringing arms from Iran to the Gaza Strip, was caught by the Israeli navy (2 January 2002), Arafat played dumb and claimed that he had no idea for whom the contents were intended. When Moshe Yaalon, then head of the IDF Intelligence Branch, complained in 1995 about the continued activities of Muhammad Def, master-terrorist, Arafat asked, "Muhammad who?" – while his advisors guffawed in the background.[49]

This sort of activity is understandable in light of Arafat's colorful personality and the image that he nurtured for decades.[50] But these techniques were adopted by his opponents in Hamas as well. When a Palestinian bomb factory exploded in the Sheikh Radwan civilian neighborhood in the Gaza Strip (April 1995), the organization's spokesman announced that the explosion had been caused by a child with a suitcase bomb, sent by Israel. Later he claimed that senior officials in the General Security Service (GSS) had initiated the attack.[51] When a suicide bomber blew himself up in a wagon bomb in the Katif area in the Gaza Strip (June 1995), Hamas announced that they had nothing to do with it, and insisted that the person had acted on his own.[52] This claim made use of the general ignorance of the neutral audience (and of Israel's failure to reach this audience) when it came to understanding the complexity of planning and executing such an attack: finding the potential suicide bomber, recruiting and training him, gathering intelligence on the target and the approaches to it, not to mention everything connected to the explosives themselves, obtaining them, preparing them, and bringing them to the chosen target.

One can see the cumulative effect of encouraging despair in the opponent – successful repulsion of any accusations by way of accusing Israel with every sin in the book, and causing reverberations in the neutral target audience favorable to the Palestinian version of events, as absurd as it may seem.

## Common Palestinian PSYOP Techniques

### 1. Creating and Exploiting Opportunities

These are two basic – and venerable – concepts. In the first case, an event may be staged from beginning to end; in the second, one makes use of an event that has occurred for PR purposes. In both cases, the initiator wishes to make use of an opportunity in order to further his goals.

By far the most successful example of creation of an opportunity was the case of the boy Muhammad a-Dura. Talal Abu Rahmeh, a Palestinian photographer for the French television network FRANCE 2, documented the boy being shot to death, the bullet coming from an IDF post at the Netzarim junction in the Gaza Strip (30 September 2000 – the second day of the Al-Aqsa Intifada). In retrospect, it is quite evident that the Palestinians staged the incident from beginning to end. The video clip that was presented includes exchanges of fire, and then shows the boy lying on the ground. The middle part is missing, and a French court criticized the network severely for this presentation of events. This "incident" contributed much to inflaming the Palestinians and was one of the causes of the escalation of violence against the IDF. It was used for years afterwards in order to encourage hatred of Israel among the Palestinians, to recruit suicide bombers, and to demonize Israel in the international arena.

The Palestinians made use of the ongoing events and of the Israeli enemy's lack of understanding of the role of PSYOP in low-intensity conflicts. Gradually, they became skilled in finding in any event or story the angle that would serve to demonize Israel. Every physical injury to a Palestinian became "genocide", "ethnic cleansing", and "a religious campaign against all Muslims". Every routine IDF patrol became an "invasion". Israeli statements were taken out of context,[53] and any Israeli stance was presented as a violation of an agreement.

Among such incidents was the riots following the opening of the Western Wall Tunnels to tourists (September 1996). This was presented by the Palestinians as a Jewish plot to take over the Temple Mount, and served to arouse the Arab public and the entire Arab and Muslim world against Israel. This principle was implemented again in connection with an incident in Gaza (June 2006), when a local family went for a walk on the beach, and several members were killed by an explosive charge. The Palestinians accused Israel of deliberately firing an artillery shell from a navy boat towards the family.[54] A number of weeks later, one of the family's daughters said, from her bed in an Israeli hospital, that her father had been playing with an unexploded shell he had found on the beach.

The construction of the security fence between Israel and Judea and Samaria, begun in April 2002, was used by the Palestinians as a pretext for a number of successful PSYOP activities. They worked to change the fence's route; complained of the damage caused to the local population, and presented it as ineffective (it can be traversed using a rope ladder). The wall was dubbed the "Apartheid Wall", and Israel's explanations that it began to build the wall in order to prevent suicide attacks coming from the PA were ridiculed.

## 2. Arousing Emotions

It is well known that in persuasion, the use of emotions can be much more effective in achieving one's goals (especially in the short term) than attempts to encourage the consumer/target audience to consider the facts rationally. The Palestinians applied this principle in their PSYOP and refined their application of it to an art.

While during the World War I countries used the arousal of emotions for political and military purposes, the Palestinians are particularly notable for their ability to produce and plan feelings such as rage, enthusiasm, and hate on demand – and this in ever increasing intensity over the past four decades. Beginning with the First Intifada (1987–1991) their ability to channel and organize rage was extraordinary, in events from neighborhood demonstrations to mass rallies.

When it became necessary to make a rational and reasonable presentation, such as in an interview with the Western media or in meetings with senior Western political figures, the Palestinians used facts and rational reasoning laced with emotional themes.

Much was made of the "evils of occupation" and messages were disseminated about the disruption of everyday life caused by the territory closures, the security fence, and the roadblocks. A picture of a Palestinian child standing on a bulldozer trying to convince the operator not to wreck his house, and a picture of a girl who lost an eye after being hit by a rubber bullet, are characteristic of this type of operation.[55] This venerable principle was taken to extremes in the Second Intifada, with its powerful and memorable images of the young boy Muhammad a-Dura in his father's arms (see above) and Sheikh Yassin in a wheelchair.

These and other images aroused internal arguments in Israel – as the Palestinians expected them to – and limited the options of the Israeli security forces. The internal pressure on the government of Israel was complemented by pressure from European countries and from the US, exhorting Israel to restrain its attacks on the Palestinians.

## 3. Overburdening the Enemy's Systems

It does not take much effort to create an overload on various systems in Israel, such as the emergency crews, police sappers, and the IDF. The goal in overloading these services is to cause them unnecessary expense, and make it difficult for them to function. An overload creates the impression of government incompetence and this was precisely the purpose of the Palestinians.

This principle was fully realized in the legal arena during the First Intifada (from the beginning of 1988), when the United Leadership worked to overburden the military court system in Judea, Samaria, and Gaza in order to prevent Israel from presenting a façade of democratic justice. During this period, a wide range of tactics was used: false alarms for the medical emergency crews, dummy explosives for the bomb squad, and false information to the police that caused the closure of roads – a nerve-wracking experience for drivers. Another effective tool was announcements by Hamas promising more painful attacks.[56]

Such simple announcements, broadcast on the radio or passed on to news agencies, raised the alert level of the IDF, the police, and the GSS. Leaves were cancelled, causing a chain reaction of tension and nervousness that affected the families of the soldiers and police officers, and then the entire Israeli public.

## 4. Media and PR

From the PSYOP viewpoint, the media are an important tool of dissemination, since except for the commercials, the content they produce is thought of as being objective by most target audiences. The Palestinians who deal with the media are successful in their competition against thousands of other groups that send material in the hopes that it will be published or broadcast: the Palestinian news in many cases deals with life-and-death situations or other dramatic issues, making it more attractive.

There is also strong competition between the various media networks, which means that moral or patriotic considerations tend to be marginalized, and messages which are from the enemy, or videos of executions, are published or broadcast. The unending demand for good stories causes items to be passed on from network to network and from one medium to another, each time with a seemingly original addition – new details or a new angle.

Events organized for the benefit of the media are known from early history of PR. It is easy to organize a press conference, but that in itself is not very dramatic, and will therefore usually not receive much coverage. On the other hand, interviews with reporters on the spot, especially immediately following an event, have much greater dramatic value, and in every war or conflict there are many opportunities for such interviews. The PSYOP crew will seek various stories and angles that will interest the media in order to pass on themes and messages that they wish to disseminate among the target audiences.

**Media management** Media management is a separate discipline in communications research, and much effort has been invested in researching the media during wartime. PSYOP, from its practical perspective, sees the media merely as a tool to achieve a goal.

In wartime, the media is far too important to be allowed to operate freely. Therefore, in totalitarian regimes – the PLO and the PA included – the media is placed under strict supervision. Democracies, on the other hand, seek to control the media without seeming totalitarian, and as a result the supervision has to be much looser.

The PLO developed media contacts beginning in the 1970s, and taking a long-term outlook, created a network of press agencies and provided them with good, usable material: translations, interviewees who spoke various languages, field trips, information, good stories, and authentic hospitality.

Seeing that the Israelis relate negatively to the Palestinian media, the Palestinians made sure to maintain good relations with Israeli and foreign media organizations and journalists, understanding that they are the most important channels of communication to the Israeli and neutral target audiences. Indeed, the Israeli media was one of the main means of dissemination of PA material. The Palestinians invested much energy in creating good relationships with the territories' reporters: they granted meetings with Arafat and other senior officials, provided generous hospitality and gave out information on various events that were interesting to the media – information that was generally reliable, though of course everything was presented from the Palestinian viewpoint. After Israeli citizens were forbidden to enter the territories, the Israeli reporters' dependence on the Palestinians for information only increased.

**Surveys** Surveys are an integral part of PR, for numbers and statistics always bear an aura of credibility. Few can analyze survey data objectively, and there are many different subjective ways to do so; in addition, the results of a survey can be biased from the beginning by tendentious questions.[57]

The Palestinians learned to use statistical data in order to strengthen their own credibility.[58] A number of research institutes were established in Gaza, Judea, and Samaria after the Oslo Accords, some of them with economic objectives and others for political purposes. Veteran research institutes such as PASSIA, headed by Dr. Mahedi Abed el-Hadi, a veteran PLO figure, and Ghassan Hatib's JMCC, used surveys in many cases. The best known institute is The Palestinian Center for Policy and Survey Research, headed by Dr. Khalil Shikaki.[59]

## Conclusions

Assessing effectiveness has always been the Achilles' heel of PSYOP. There are few people or bodies (either private or public) that will readily admit that they have been influenced by direct or indirect messages aimed at them. And decision makers, of course, are even less likely to make such an admission. As this discipline has become more professional, methodical tools, such as polls, surveys, and other studies of the behavior of the target audience, have been created to assess PSYOP effectiveness.

The data on PSYOPs during the period under discussion here, if there are any, are apparently being kept secret. So, lacking hard numbers, we can only take a panoramic view of the conflict and its conditions and offer an overall perspective that categorizes the enormous number of messages sent according to the operational PSYOP principles utilized.

As mentioned, there is no available information (and possibly no secret information either) on Palestinian PSYOP activity, such as a manual for Palestinian PSYOP. But an analysis of Palestinian activities according to PSYOP principles shows that the Palestinian messages were not created and disseminated thoughtlessly and spontaneously, but rather the opposite, and that they were designed to serve some very well-defined goals.

This chapter suggests that the operational paralysis that took hold of Israel during the entire Oslo period and until the middle of the Second Intifada was a direct result of the successful strategic PSYOP campaign of the Palestinians, which was combined with violence: rockets were fired on Israeli cities, suicide bombers exploded in its streets, Israel's international status was eroded, and the nerves of its citizens were sorely tried. A temporary change in this situation occurred not because of an Israeli initiative, but because of a Palestinian mistake of yanking the chain too hard. The suicide bombing in the Park Hotel in March 2002, which Israel wisely labeled the "Passover Seder Massacre", was too strong a tug (as was also the kidnapping of Regev and Goldwasser by Nasrallah four years later), and Israel embarked on a limited military campaign. And yet – though one cannot deny Israel's military achievements – as in other cases, the Palestinian PSYOP managed to gain control of the political situation almost immediately, by rebranding the Israeli offensive as the "Jenin Massacre".

In retrospect, one has to admire the Palestinian achievements, considering the asymmetry between them and Israel. In spite of the fact that the Palestinian PSYOP often seems to be primitive and unpolished, and even to contain some Goebbelsian elements, what matters is the end result. No deep research is needed in order to give statistical proof of the erosion in Israelis' morale: it is enough to look at the data regarding emigration of Israelis to other countries, the number of foreign passports Israelis acquire, and the number of residential real estate deals they have been making abroad.

## FIFTEEN YEARS OF PALESTINIAN PSYCHOLOGICAL WARFARE

Palestinian PSYOP began to bear fruit during the First Lebanese War in 1982. From faltering beginnings the Palestinians built up, five years later, an impressive system of information dissemination to various target audiences – home, enemy (Israel), and neutrals – which had its effect during the First Intifada: the very fact that "Intifada" became a household word around the world is an indication of their success. This system generated, made use of, and marketed information using such simple means such as leaflets and faxes, but to great effect, leading to its first significant political achievement: the Madrid Conference (1991). In the following two years the Palestinians focused on secret negotiations, which brought about their next achievement, the Oslo Accords. These Accords created a de facto state and allowed the PA to upgrade its operations in many ways; in particular, its PSYOP became both open and official. It was used in order to continually improve the Accord's conditions, and PSYOP became an integral part of negotiations. Early on in the conflict, Arafat focused his diplomatic efforts on the West in order to convince leaders that they should negotiate with him; his success in that area is truly astounding.[60] And yet, in retrospect we see that Arafat never gave up on the idea of a Palestinian state "from the river to the sea", and that on the covert level he continued to work towards the dismantling of the State of Israel. Indeed, when the celebrations and the signing ceremonies were over, Arafat and Hamas moved on to guerilla warfare and terror. So once again, classical PSYOP doctrine takes center place as a shaper of political realities.

Arafat's death and the split in Palestinian society between Hamas and the PA did indeed slow Palestinian progress, but while the situation is in constant flux, the balance of power does not change radically. The goal remains to have PSYOP serve the Palestinians' military and political ends, to cause the Israelis despair and embarrassment, and to divide Israeli society, convincing it that justice demands a correction of an historical wrong. And, of course, if they only accept this one current demand, everything will work out.

The demonstrations against the fence in the village of Bil'in, the academic boycott in the UK, and the Swedish organ harvesting libel are classical cases of PSYOP.[61] The idea of harvesting organs was originally a Soviet PSYOP campaign, meant to blacken the United States' image in the final years before the collapse of the Soviet Union.[62] All these things show that in between waves of military activity, there is constant Palestinian PSYOP aimed at carefully selected target audiences.

From the historical point of view, there is nothing really new in the Palestinian PSYOP, except for the suicide bombers. Everything else has been done in the past, and is known to any researcher of wars for independence, such as the struggles of the FLN in Algiers, and of the IRA in Northern Ireland. The good news – and the bad news – is that all the rules and techniques are known in advance. The planners on both sides have only to insert the variables into the equation and receive the future form of the opposing PSYOP campaign. Security forces would do well to allot suitable resources to research in this realm, which would enable them to develop a PSYOP map for the future. The only surprise would be the topics that the enemy might choose to activate with existing techniques.

The current information situation, as far as Israel is concerned, is not very helpful. There is almost no passing of Arabic messages to the Palestinians, and the response to any suggestion of setting up an Israeli satellite network in Arabic or in Farsi is negative, even derogatory. The various Palestinian groups and the Iranians continue to develop and strengthen their means of dissemination and engage in ever-growing PSYOP campaigns (the "poisoning" of Gaza, theme,[63] the Goldstone Report, nuclear-free Middle East, etc.). Israel should sabotage these moves, and reverse the Jihadization process that the PA has put the younger generation through during the period under discussion; it should initiate preventive PSYOP activity that will bring all the Palestinians to a reality-based outlook.

## Appendix A: Main Events in the Israeli–Palestinian Conflict, 1993–2008

September 9, 1993 – Prime Minister Yitchak Rabin and PLO Chairman Yassir Arafat exchange mutual recognition letters.

September 13, 1993 – Festive ceremony at the White House celebrating the signing the "Declaration of Principles on Interim Self-Government Arrangements", known as "The Oslo Accords", or "Gaza and Jericho First".

September 1995 – The signing of the "Second Oslo Agreement", which dealt mainly with the transfer of Palestinian cities in Judea and Samaria to Palestinian rule.

September 24–27, 1996 – Riots in Judea and Samaria following Prime Minister Binyamin Netanyahu's opening of the Western Wall Tunnels to tourists.

January 15, 1997 – Signing of the "Hevron Agreement" between Israel and the PA on the withdrawal of the IDF from most of the city.

October 23, 1998 – Prime Minster Binyamin Netanyahu and PA Chairman Yassir Arafat sign the Wye Accords, which dealt with practical measures to implement the Oslo Accords.

July 12–25, 2000 – The failure of talks between Prime Minister Ehud Barak and Yassir Arafat at Camp David, with President Clinton mediating.

September 28, 2000 – The beginning of the Al-Aqsa Intifada.

March 29–May 10, 2002 – Operation Defensive Shield, in the course of which the IDF entered the Palestinian cities in Judea and Samaria.

November 11, 2004 – Yassir Arafat dies.

January 15, 2005 – Abu Maazen is appointed Chairman of the PLO and second President of the Palestinian Authority.

Summer 2005 – The Al-Aqsa Intifada ends, for practical purposes.

August 15–23, 2005 – Execution of the disengagement plan, in the course of which Israel evacuated all the Jewish settlements in the Gaza Strip, and four settlements in northern Samaria.

January 25, 2006 – Hamas is victorious in the Palestinian Parliament elections.

March 29, 2006 – The Hamas government is sworn in, headed by Ismail Haniyeh.

June 11–16, 2007 – The Gaza Strip is taken over by Hamas.

December 27, 2008 – The beginning of Operation Cast Lead in the Gaza Strip, following the launching of dozens of Kassam rockets and mortar shells into Israel after the end of the six-month-long Tahadiyah.

# 4

# Hasbara, Propaganda and Israeli Public Diplomacy

## A Historical Perspective

An important consensus in Israeli public discourse is that "the Israeli national image is a disaster". This adage has been accepted as a truism for at least four decades. The failure of the State of Israel to cultivate its public image, especially in everything related to the Arab-Israeli conflict, is a phenomenon that requires attention. The failure is especially striking when one considers that the State of Israel has far more resources than its Palestinian opponent. How is it that a young country that managed to develop nuclear weapons for its defense, created a reputable intelligence service, and built up an excellent army, is so helpless in the important strategic realm of hasbara,[1] that is, in the realm of international image-building?

Numerous critics, Israelis, Jews, and pro–Zionists, have supplied a long list of explanations, and they are mostly correct. They include organizational reasons, carelessness, prioritizing problems, isolationism, and many more. One important factor is the (justified) drawbacks of being a democracy. Freedom of movement in Israel enables reporters to move all over the country, including the territories (except for the temporarily declared areas known as "closed military zone"), something that is not possible in the neighboring Arab countries. There is almost no censorship in Israel, and reporters are almost completely free to report what they want. Israel was for many years the only place in the region with the satellite technological infrastructure for broadcasting information collected in the territories to networks abroad.[2]

The freedom of opinion in Israel prevents the government from maintaining a steady political line as Arafat used to do, first as head of the PLO and then as head of the Palestinian Authority. In addition, Israel has a liberal democratic tradition and as a result the Israeli people tend to react with revulsion to what is perceived as propaganda, and associates it with lies and disinformation. That is the reason the authorities prefer to deal only with the "clean" side of hasbara, and to leave the "dirty" side to the secret services.[3] The Palestinians, by contrast, have no such inhibitions and they take every opportunity to make false accusations and stage events in order to demonize Israel.

As was said above, all these claims are true, and yet they do not give a full and convincing explanation of Israel's four-decade failure when it comes to propaganda.

The roots of this failure usually lie in both organizational factors, such as faults in the division of responsibilities, allocating resources, decision-reaching processes, professional appointments etc., and in structural factors such as cultural patterns. In the Israeli context, one can point to at least three such patterns: the traditional Jewish cultural pattern, the Israeli-Zionist pattern, and the Western-democratic pattern. (Of course, there are close connections between these three, but for the purpose of theoretical analysis we must distinguish between them.) I will focus mainly on the first two patterns – the Jewish and the Israeli – and the Palestinian approach as well.

I wish to prove that the roots of Israel's inadequacy with regard to hasbara can be found in fundamental patterns of Israeli thought. It is quite clear that some of them can be overcome, and some can even be gotten rid of, but everything has a price.

### Jews and Hasbara

In order to trace the roots of the Jewish attitude towards hasbara, one has to delve into the forces that shaped Jewish awareness of the issue. Of course, I will not be able to recount the entire history of Jewish hasbara, but I will attempt to present here some of the most important episodes, episodes that in my opinion can teach us something about the early problems of Jewish hasbara, from which its more recent faults developed.

Already in the Bible we find a Jewish sensitivity to considerations of image. When Shimon and Levi attacked Chamor and his family, in revenge for what Shechem the son of Chamor did to their sister Dina, Yaakov did not express any moral outrage about the actual act (this outrage expressed itself mainly in the blessings he gave before he died, see Genesis 49:5), but did express fear regarding the damage to his image: " . . . to make me odious in the eyes of the inhabitants of the land" (ibid., 34:30). Indeed Yaakov, already a third generation in the Land of Israel, still saw the Canaanites as the "inhabitants of the land" and himself as one who was dependent on them, to one degree or another. His image policy was mainly defensive. When Moses stood before G-d, he did not neglect the matter of image, and asked: "Why should the Egyptians say . . . " (Exodus 32:12). We see here policy (and the fate of the Jewish people) being influenced by what is known today as public opinion.

In the period following the conquering of the Land of Israel, the Jewish people faced a long series of wars and struggles in which they exhibited considerable creativeness and ingenuity, but they did not shine in the realm of hasbara. The most brilliant propaganda moves at the end of the Israelite period actually belonged to their enemies: these were Ravshake's speech at the height of the siege on Jerusalem " thou trustest upon the staff of this bruised reed, upon Egypt" (Kings II 18: 19-37) and Haman's manipulative conversation with Ahaseurus in which he convinced Ahaseurus to agree to the extermination of the Jewish people. There is a certain people scattered abroad and dispersed among the people in all the provinces of thy kingdom; and their laws [are] diverse from all people; neither keep they the king's laws: therefore it [is] not for the king's profit to suffer them (Esther 3:8).

The relationship between Jews and non-Jews during the Hellenistic and Roman periods was extremely tense, even on the plane of ideas. Prof. Menachem Stern, in his comprehensive book *Greek and Latin Authors on Jews and Judaism*, gives us a fascinating picture of the Jews' image in the eyes of non-Jews[4]. Stern's book brings scores of quotations from the writings of various authors from different periods, and the vast majority show that the Jews' image in the classical world was poor, and usually because of distortions of the facts. We have almost no sign of any Jewish effort to combat these images. Philo (20 BC–AD 50) did make an effort to explain Judaism in Greek philosophical terms, but it seems that his life work reflects not so much an effort to improve the Jewish image as a sincere internalization of this culture by Jews of his type. The translation of the Bible into Greek helped disseminate Judaism's messages, but the initiative for making the translation evidently came from non-Jews, and the Talmudic sages viewed this development almost like an evil decree.

The person who may be regarded as the pioneer of Jewish hasbara in ancient times is Josephus Flavius (37– c.100), in his book *Against Apion*.[5] This book was aimed at the Hellenistic author Apion, who had written against Jews and Judaism in order to prove that the Jewish people was a young nation (a shameful characteristic, according to the concept of the times), and that its religious customs were wrong and indecent. Josephus wished to prove the opposite and cited a wealth of sources to show that the Jewish people was an ancient people, that early Greek historians mentioned them, and that their laws were good and proper. This is not the place to judge Josephus, a complex figure who has fascinated many historians. However, from the hasbara point of view Josephus already showed all the weaknesses of later Jewish hasbara: his claims were almost entirely apologetic; there was no criticism of the opponent, except where the opponent made unfounded criticisms of Judaism; and there was no condemnation of the other side's values or culture; in fact he tried to defend Judaism by showing that it fit in with these values and culture. Lastly, there was no "wickedness" or manipulation of the facts: Josephus did not utilize the reader's prejudices, his dark inclinations, or his socio-cultural biases; he only presented fine, practical, business-like, clear, and "politically-correct" claims. Joseph's hasbara is characterized mostly by defensiveness.

The Talmudic sages held various debates with the wise men of the non-Jewish world – "philosophers", "matrons", and others – and even with opposing sects within the Jewish world. These debates were very different from those that Josephus held, but nevertheless they also show little respect for the opponent and indeed a conviction of the impossibility of convincing him of the truth of Judaism. In many places in the Talmud and contemporary literature (Midrash) we find that after the Sages answered their opponents, their students came to them and asked: "That one you pushed off with a reed – but what answer would you give us?"[6]

The implicit assumption is that the opponent is not worthy of a real answer, and that even if given one, he would not understand it. But it also seems to be assumed that the real answer might touch on some raw nerves, and we should not be in a hurry to do that. In private discussions among the Sages we sometimes see hatred towards the non-Jews, the Sadducees, the *minim*, the heretics, the ignoramuses, and other "others", but in public debates with them the Sages' attitude was always one of careful politeness, and sometimes even apologetic. It seems that their idea was the same as that lurking behind the well-known adage of R. Eliezer: "Work hard on learning Torah, and know what to answer the heretics" (Avot 2:14): you must know the Torah, and you must also learn to present it and defend it. There is a concern here for the image of Judaism in the eyes of the non-Jews, and perhaps also concern for the image of Judaism in the eyes of the "audience" of the debates – society in general – but at the same time there is despair regarding the possibility of convincing others to see matters through Jewish eyes. The debates with the *minim* – especially with the early Christians – also seem to reflect little respect for others' opinions, and it is clear that the purpose of the debates was to check the spread of their message to even larger segments of the population.

On the face of it, entering into debates with other groups and with schismatics within Judaism shows basic respect for their opinions. Indeed, we find in the Talmud (Shabbat 33b) a statement of Rabbi Yehuda (second century AD) about Rome: "How nice are the deeds of this nation: They have built markets, built bridges, built bathhouses." On the other hand, the attitude of Rabbi Shimon Bar Yochai (second century AD) toward Rome is quite negative: "Everything they did, they did only for themselves: they built markets – to sit prostitutes in them, bathhouses – to pamper themselves, bridges – to collect the toll." (ibid., and see also Avoda Zara 2:b). The debate reflects two different perceptions of the conquering superpower: according to one view, Rome had a certain amount of moral power ("how nice are the deeds") as well as physical power; while according to the other, the deeds of the superpower were devoid of virtue ("Everything they did, they did only for themselves").[7]

Each of these perceptions produces a different view regarding the value of ideological debate with the Romans: the one perception suggests that there is room for an intercultural dialogue between Judaism and the classical world, while the other suggests that there is no room for any such dialogue. It should be noted that Rabbi Shimon bar Yochai was Rabbi Akiva's disciple. Rabbi Akiva served as the spiritual leader of the rebels against Roman rule, and supported the idea of an active struggle against the conquering empire. It is true that Rabbi Akiva also wrote another dialogue between religions, between Jews and "wicked Turnus Rufus" (Bava Batra 10a), rather than with philosophers. However, that dialogue between captor and the captured is not a real, free, and open dialogue, and one can see that Rabbi Akiva did not really expect to convince his opponent; and in fact, as mentioned above, he went on to pursue a military strategy. This same approach was expressed one thousand and eight hundred years later by Ben-Gurion.

We find a similar approach in the debates between Jews and Christians during the Middle Ages.[8] Here too, most of the debates were forced on the Jews, and a favorable image was presented not out of a desire to convince the opponent of the truth of Judaism, but out of a desire to survive.[9] There was never any strident criticism of the opponent – the Christians – who were massacring and persecuting others in the name of love and mercy, nor was there any criticism of the values of Christian culture in general. The debates always focused on exegesis of the scriptures and on theological tenets. It seems that during the Middle Ages there was a general opposition to Jewish missionary activity, while in the times of the talmudic sages there was at least some advocacy of such activity.[10]

In an earlier period of the Middle Ages there was also a fierce debate between the Jews and the Karaites.[11] This debate produced a large body of literature aimed at proving the truth of rabbinic Judaism, and it included some of the masterpieces of Jewish thought.[12] This intellectual struggle was accompanied by a struggle for political power, about which we do not have much information. But it seems that at this stage the option of hasbara had a clear advantage over violent action. Maimonides said that one must distinguish between the ideologues of religious perversion and "the sons of these mistaken people and their grandsons, whose parents misled them and they were born amongst Karaites and raised according to their opinions." While the former are in the category of those "to be sent down" (in other words, full active measures may be taken to do away with them), each person in the latter group is like "a child who has been taken captive" and rates a different treatment: "One must cause them to repent and attract them with peaceable words until they return to the true Torah." In other words, one must use hasbara with them.[13]

With the elimination of the walls of the ghetto, in the eighteenth and nineteenth centuries, Judaism faced a new front. The most notable representative of Jewish hasbara at the time was Moses Mendelsohn (1729–1786).[14]

In some ways, Mendesohn's work resembles Josephus' *Against Apion*, and while Josephus did what he did out of self-defense, Mendelsohn was fully aware of what he was doing. The most notable example of this is the famous "Lavater controversy". Lavater, a priest, challenged Mendelsohn to defend the tenets of his religion in a public debate or convert to Christianity. Mendelsohn was afraid, for if he were to enter into a religious debate it would contradict his belief in bringing Jews and non-Jews together and of cultivating mutual respect between the religions; if he were not to engage in the debate, he would seem weak and helpless in the face of his opponent's challenge. He chose to write a letter titled "Jerusalem" in which he explained why he thought that he should not participate in the debate.[15] He was convinced of the truth of his religion, and would not have accepted it without having sufficient reasons for it. However he was prevented from presenting his reasons, both because of the Jews' low status in Germany and because he believed that in the end Christianity's values are effective in perfecting mankind, even if these values derive from theological principles that he thought were mistaken. He therefore thought that it was better to leave the Christian with his faith than to undermine it.

Mendelsohn did not miss the opportunity to boast that Judaism, in contrast to Christianity, is not a proselytizing religion.[16] One thousand and eight hundred years of submissive Jewish behaviour, which was not combined with any proselytizing military operation or even the threat of one, has left its mark upon the Jews. The emotional baggage deriving from the complex relationship between Jews and non-Jews affects both the hidden and the open debates about Jewish methods of hasbara. The new realities of the emancipation presented Jews in Central and Western Europe with an unprecedented challenge: some of them converted to Christianity, while others remained in a virtual ghetto because of their clothes, language, and "Jewish" professions.[17]

Between those two poles there was a full range of identities and solutions for the relationship with the non-Jewish world: reform in its various guises, a secular Jewish identity, modern Orthodoxy, anti-modern Orthodoxy, and later the Zionist national identity.

The question of the legitimacy of bringing non-Jews into the Jewish faith began to be addressed in the nineteenth century. Although there was no Jewish missionary movement, in Italy Rabbi Eliyahu ben Amozegh (1823–1900) developed a plan for setting up a "Sons of Noah" movement for non-Jews who had become convinced of the truth of Judaism.[18] He suggested that instead of converting, these non-Jews accept the seven Noachide laws, which are the ideal that Judaism offers those who were not born Jewish. Many years later this idea became popular, and even acquired considerable influence after it was adopted by R. Menachem Mendel Shneersohn of Lubavitch.[19]

Indeed, the nineteenth century was rich in apologetic literature about Jews and Judaism. The most outstanding representative of this approach was Rabbi Samson Raphael Hirsch (1808–1888), almost all of whose writings are in German, and represent an attempt to enwrap Judaism in the clothes of German Romanticism.[20] Rabbi Hirsch's ornate rhetorical style shows clearly that his writings are works of hasbara that go beyond the task of imparting ideology. The Orthodox circles in Hungary and Galicia, which called for avoiding any kind of debate or ideological conversation with the "outlaws", and held that laboring in Torah was the telling answer against assimilated and heretic Jews, conducted intensive internal campaigns in order to strengthen the ranks of traditional Judaism through scores of sermons, letters, leaflets and personal visits to communities.[21]

## Zionism and Hasbara

At the same time, this period was one in which modern, racial anti-Semitism developed and largely supplanted religious anti-Semitism.[22] The range of responses to the anti-Semitic challenge reflected the range of Jewish existence. At one extreme there was the belief that anti-Semitism would disappear if the Jews would only forsake Judaism on one level or another. At the other extreme there was the approach represented by the second century saying "It is a known law that Esau hates Jacob",[23] which held that any attempt by Jews to lessen the gap between Jew and non-Jew would only make matters worse.[24]

Within this range we also find the heads of the Zionist movement, for they were searching for a solution to the problem of anti-Semitism and wished to establish a state for the Jews. Unlike Leon Pinsker (1821–1891) who saw anti-Semitism as a disease,[25] Theodor Herzl (1860–1904), some of whose friends were anti-Semites, showed understanding of its causes and thought that anti-Semitism had a positive role in a solution of the "Jewish problem".[26]

An intensive striving for favor, recognition, and integration influenced the behavior of those who left the ghetto. After generations of rejection, they now had to overcome not only social and legal barriers, but also emotional barriers, and they were most anxious to change people's feelings towards them. They were willing to make a tremendous effort in order to achieve that goal – to publicize, to lecture, and even to change their lifestyle. One can call this phenomenon "longing for love", and it influenced the state that they would establish in the future, and also be a powerful weapon in the hands of its enemies.

The Zionist movement, which was full of internal discord in all matters relating to the relationship with other nations and the possibility of absorbing their cultures, failed to change much in this realm. The World Zionist Organization did abandon Herzl's recommendation that hasbara be a central goal of Zionist activity. However, even the supporters of Practical Zionism understood, although to a lesser degree, that the Zionist enterprise depended on the perception of its legitimacy in the non-Jewish world. Herzl's recommendation reflects an internalization of the European way of thinking, an awareness of the value of hasbara and of the importance of being aware of the cultural and moral contexts in the successful advancement of an idea.

The way of the Eastern European Zionists, who were to set the tone for the settlement of the Land of Israel, was different. David Ben-Gurion (1886–1973), who unseated Chaim Weizmann (1874–1952) and imposed a socialist outlook on the World Zionist Organization, promoted activism together with a rejection of the traditional Jewish way of thinking. In his fight against the British Mandate, Ben-Gurion used political means such as speaking to committees investigating the issue, and did not refrain from using the media. However, once Israel's independence was achieved, the activist element was emphasized more and more, and Jewish leaders devoted great attention to settlement, development of the economy, immigrant absorption, and achieving military independence. His partner and rival, Moshe Sharet, was put in charge of building Israel's foreign relations. There was perhaps not even time to invest in hasbara. The Zionists were badly disappointed when it became clear that Jews were not coming in hordes to the new state, and also that anti-Semitism had not disappeared. But they consoled themselves with the thought that the Jews would eventually come to Zion, and that the source of the anti-Semitism that had appeared in the Arab states was the Arab-Israeli conflict, and once that was solved, the hatred would disappear.

In terms of its content, Zionist hasbara was based entirely on the effort to achieve legitimacy, and the claim was and still is basically a moral one: we deserve it because we are a nation among nations; we deserve it because we have an "historical right" to the Land of Israel; and after the Holocaust, we deserve it because we were slaughtered and the option of exile does not exist anymore. All in all (except for a certain number of activities by the Etzel and Lehi underground organizations), not much effort was invested in PSYOP in order to arouse animosity, opposition, or fighting instincts against the British conqueror (to this day the average Israeli bears no grudge against the kingdom that ruled his land for over thirty years), or against the Arab enemy, even after repeated murderous attacks. The Zionist attempt to create a "new Jew" – muscular, proud, sure of himself, fierce – was aimed mainly inward, in efforts to destroy the remnants of the exile mentality;[27] outwardly, Zionist hasbara focused on the "underdog doctrine". Even later, when the entire world was amazed at the achievements of the young state – in no small part because of its military power – there was little deviation from the concentration on this doctrine, and Zionist hasbara focused on the moral supremacy of the Israel Defense Forces and on the moral right of this young society to protect itself. Again, there was almost no "PSYOPing" of the "wicked" type, aimed at blackening the enemy, weakening its beliefs, nurturing opposition within the enemy, and awakening feelings of contempt among the domestic audience for the enemies' culture, values, and strangeness. Of course, officially Israel could not engage in such methods, but anyone familiar with international strategic public relations and psychological warfare (PSYOP) knows that moves in this area are often done clandestinely by sending the necessary messages through indirect channels, camouflaging the true source. It is possible that the need to prevent harm to the Israeli Arabs and to promote their becoming citizens contributed to a hesitancy to engage in negative hasbara.

The country's failures in the realm of PSYOP quickly became apparent, in terms of both organization and content. Regarding organization, it is obvious that there was a lack of clarity about the goals of the hasbara, its audience, the methods that needed to be used, and its importance (expressed, of course, by budgetary allocations), which resulted in shifting the responsibilities for hasbara from one body to another. The authority for it was divided up according to political and personal connections rather than according to professional ability.

The basic assumption of Israeli hasbara was that the message must reflect the Israeli consensus. Beyond the Zionist consensus about the importance of a Jewish state (shared by most sectors of society other than the far left and the ultra-Orthodox), there was no agreement about the desired character of the Jewish state and the course of action it should take with its enemies. The Jews brought their ideological split in the attitudes towards non-Jews, caused by the opening up of the ghetto, to their discourse about the policies of their state; and they raised the question, based mainly on an apologetic approach: What is the moral justification for expelling people, some of whom have been inhabiting the land for generations? The internal argument in Israel was not unknown to international observers, and it goes without saying that it strengthened Israel's enemies in their political campaigns against Israel. This moral debate became a disadvantage for Israeli hasbara abroad. On the psychological level it is clear that a careful, hesitant approach undermines the professional ability to build a dynamic and exciting campaign. Moreover, the earlier goals were not relevant anymore, for Israel's hasbara struggles in the world were no longer about its right to exist, but rather about what it did as an existing state. On those issues, as mentioned above, there was no consensus even in the Israeli public.

But the worst failure of Israeli PSYOP efforts was in relation to the target audiences. Almost all resources were dedicated to the "Poretz" (non-Jewish landlord) abroad, primarily the US and Europe, and there was no attempt to influence the Arab Palestinian adversary that Israel was facing. The Jews tended to assume that the basic beliefs and opinions of the adversary, and his adherence to his culture and values, could not be undermined. There was almost no effort to emphasize or exploit the violent rifts between Christians and Muslims, Sunnis and Shi'ites, and the religious and the secular within the Arab movement. No attempt was made to remind the Palestinians that their struggle was, after all, about land, and to present them with the dilemma of "land vs. blood". The average Palestinian was not confronted with shocking images of the results of Palestinian terror attacks, while Israelis saw photograph after photograph showing the suffering of the Palestinians. Until the past few years, no attempt was made to reveal to the average Arab-on-the-street the corruption of his leaders, who are quick to send the average family's sons to the killing fields but keep their own sons close or in a safe place abroad. Even in the past few years, when Israel has addressed the matter of corruption within the PA, it was done mainly to convince the powers that be in the US, and less as part of a PSYOP campaign directed at the Palestinians. Moreover, even when the religious factor motivating Palestinian terror became stronger, Israeli psychological warfare (PSYOP) did not utilize any themes of "religious war" which succeeded so well in Christian Europe, such as exposing the corruption, lies, and hypocrisy of the Palestinian religious leadership. This was an important missed opportunity, for the state had enormous manpower resources in the immigrants from Arab countries, who were familiar with that culture to the point of intimacy, and could have shaped and helped deliver the necessary messages. But the cultural and managerial power in Israel was in the hands of people of European descent,[28] and the hasbara failures can be seen as reflecting the historical struggle for dominance in the Jewish world between Sephardi and Ashkenazi Jews.

In short, Israeli hasbara has remained entirely defensive, and aimed entirely at public opinion in faraway lands that were not involved in the conflict.

## The Palestinian Opponent, its Hasbara, and its Impact on Israel

In contrast to the repeated historical failures of the Jews to develop effective hasbara campaigns, the Palestinians have never stopped their hasbara activities. In the initial phases during the 1950s, the Palestinians were influenced by a number of different revolutionary ideologies: communism, socialism, Maoism, etc. But their primary objective was clear. Like the Jewish Yishuv before the establishment of the state, the Palestinian leadership was less concerned with the character of the state after independence than with ways to achieve independence. When the power of the Israeli military was demonstrated time after time, the PLO understood that guerilla tactics were irrelevant, and they turned more and more towards strategic political public relations.

As discussed in the preceding chapter, three central themes may be discerned in the Palestinian PSYOP messages of the past thirty years. The first is "from asset to liability", the second is guilt, and the third is justice.

The "asset to liability" theme aims to convince target audiences in the enemy population (and also neutral observers) that the benefits to be derived from the continuation of the occupation are fewer and smaller than the losses it entails. These losses include lives, political prestige, economic damage, and so on.

The second theme is guilt. One must cause the enemy to feel guilty. As Ellul noted in his monumental work from the 1950s, an army that feels guilty has lost its efficacy entirely.[29]

Creating guilty feelings is a well-known psychological warfare technique, and its goal is to cause the enemy soldier to stop and think and question the beliefs instilled in him by his state in order to convince him to commit violent acts that he would never do as a humane, law-abiding person. The Palestinians did this by publicizing their difficult living conditions and the sacrifices of their civilian population, and by emphasizing the image of the "Zionist soldier". These images were designed to influence not only world public opinion, but also Israeli public opinion and Palestinians' self-perception as well. The Palestinians broadcast a remarkably consistent and cohesive message.

The third theme is justice, and that was aimed mainly at neutral audiences, and to a lesser degree at the Israelis. According to this theme, the Palestinians deserve a state because of their suffering and their historical rights. Much effort was invested in establishing a plausible basis for these rights.

It was not difficult for the Palestinian message to influence Israeli public opinion. In contrast to Arab societies where information is filtered from above by strict censorship and from below by the self-discipline of the media, the Israeli media were open to accept messages originating from the enemy. Thus the Israeli media operated according to a self-damaging double standard: harsh pictures of Israeli victims of terror were not broadcast due to respect for the victims, but harsh pictures of damage done to the Palestinian population were broadcast, sometimes at great length and in great detail.

The Israeli public could not withstand the pressure. Israelis who were asked to explain the country's behavior, and to show how it could be acceptable to world public opinion, found it difficult to do so. Thus the Israelis' moral strength and power of resistance were impaired.

There is no doubt that the success of the Palestinian campaign was made possible to a large extent by socio-cultural developments within Israeli society. Israeli society, founded on a socialistic-Zionistic ethos, went through profound changes in a short time. The young Israeli society had placed great emphasis on the collective identity, and many social systems were created in order to nurture the "new Jew" of the Zionists. The generation of the 1940s wanted to return to the times of the Bible, and it saw itself as fighting for national liberation. But with the rise in the standard of living and level of consumption – and as Israel became a Western-liberal-capitalistic society – the values of the "age of ideologies" were pushed aside and replaced with the values of freedom of the individual and his or her right to self-fulfillment. The rulings of the Supreme Court of Israel during this period not only clearly reflect this change, but also actively furthered it. The Bible got pushed to the periphery – especially among secular people – and since the ultra-Orthodox had never made it the center of attention anyway, the only group that continued to fashion itself according to the Bible was the National-Religious camp. As the state became better established and stronger, wars stopped being seen as wars of national independence, and the siege-like feeling decreased. Moreover, as the 1940s generation gradually withdrew from the political and cultural stage, it became convenient to dim the myth surrounding it, particularly for the new historians, who began as a small and eccentric group of far left-wing representatives and proceeded to penetrate to the heart of Israeli academia, and even disseminated their messages in academic forums in the West.

In this atmosphere it was easy for the Palestinian narrative to be accepted and thrive. The Palestinian strategy in this regard was very much influenced by the propaganda doctrines of the communist revolutionaries: first one should convince the revolutionary avant-garde in the cultural elite; this elite will internalize the message and identify with it; at the beginning this group will seem rejected and deviant, but it will serve as the core from which the message will spread to ever wider circles that initially do not identify with it. Circles that do not identify with the message will nonetheless receive it in "translation", in a smaller and more diluted dose, so that even if they do not internalize it, it will be easier for them to get used to it, and their ability to resist it will decrease with time.

The Palestinians' use of communist propaganda techniques for inspiration resulted in a phenomenon that can be called "multi-level propaganda". This occurs when not just the form, but also the content of a message is adjusted for different target audiences, resulting in different messages for different populations. This technique was applied successfully in the Islamic republics of the Soviet Union, which at the beginning were not ready to accept Marxism-

Leninism, not even in its lighter, diluted form. It was determined that soviet propaganda directed at these republics had to base itself on traditional – and sometimes even religious – themes, occasionally arousing the anger of the soviet leadership.

Similar multi-level messaging can be seen in Arab propaganda activity, especially in matters relating to anti-Semitism. Arab anti-Semitism is a very interesting phenomenon. The fact is that the Arab world is today the main consumer of anti-Semitic publications such as the *Protocols of the Learned Elders of Zion*. The Arabs need anti-Semitism in order to strengthen their cohesion and their hostile spirit towards the Jewish state, which is seen as a foreign body in the Middle East. They also need it in order to give themselves a psychological answer to the question of how five million Jews managed to vanquish larger Arab armies chosen from a population of hundreds of millions.

This anti-Semitism is in direct contradiction to the image that the Arabs try to project to the Western world. They make two central claims in order to refute the charge of anti-Semitism: first, that Arabs have nothing against Jews, only against Zionists; and second, that it is impossible for Arabs to be anti-Semites, as they are Semites themselves. In fact these claims have nothing to do with reality. Regarding the first, one can see that Arab anti-Semitic images are aimed specifically against Jews as Jews, using the symbols of both old and new religious anti-Semitism. As to the second claim, it is merely a semantic ploy, for the term anti-Semitism is aimed only at Jews, according to the European tradition of its use. How can the Arabs generate these anti-Semitic messages while in the Western world they present a different face and refrain from anti-Semitic and racist utterances, all the while accusing Zionism and Israel of racism? The answer lies in the communist doctrine of multi-level messaging, which has deep roots in different cultures. It is not always necessary to worry about contradictions between the messages that are broadcast to different audiences, for sometimes the cultural borders between the two societies targeted do the work properly. A person watching television in New York is not interested in what message the person watching television in Cairo is receiving, and vice versa. Israel has tried for decades to expose the two-facedness of Arab messages, but to no avail: publications addressing the issue did not receive any notice whatsoever.[30] Only after September 11 did US heads of state begin to interest themselves in the messages that Arab children are being raised on, and what they found – even in the schools of ally Saudi Arabia – did not make them happy.[31]

In order to engage in propaganda, one should be completely convinced of the justice of one's way. Someone who himself has doubts will find it very difficult to convince others. In Israel there is a severe problem regarding Israeli perception of the justice of Israeli claims. The problem exists especially in regard to the question of the expulsion of the Arabs in the War of Independence (1948).[32] The Israeli feeling of uneasiness about this matter stems not so much from familiarity with the facts, but rather from a lack of familiarity with the facts. As Joel Fishman has shown, there is a prevalence of lack of attention to historical knowledge in Israel.[33]

This general Israeli lack of historical knowledge stems from several causes, and with regard to the War of Independence in particular, it stems from some very specific factors. During the war, many of the documents in military headquarters were not preserved, and since the Israel Defense Forces were just beginning to get organized, there was great disorder in their camps. In addition, officers made sure to cover up failures, especially since many of them later entered politics. In everything connected with the Arab side, there was also the fear of political sensitivities, and it seems that not a few documents were destroyed or classified. Another factor which has caused the Israeli people to neglect their nation's history is the anti-intellectual process that Israeli society underwent as the standard of living rose. The fine arts were rejected in favor of practical arts, and the humanities were deeply affected by this trend. Thus the New Historians could rewrite history and have it accepted by the generation that grew up some decades after the war.

The Palestinians, on the other hand, devote great attention to history and documentation; in particular the PA has allotted great resources to this end. The Orient House, headed by Faisal Husseini, was the site of extensive activity in this realm. The Palestinians certainly engage in blurring and covering up, but one can see that their tendency is to make research subservient to propaganda interests, and not to have it work against them. As part of this tendency, the Palestinians have developed new historical arguments even about the distant past. For example, an argument was developed claiming that the events of the Bible took place in Africa, based on linguistic and archeological research. There is also an argument that presents the Hebrews as a nomadic tribe that invaded and integrated with the local populations, the Palestinians being the direct descendants of the Canaanites.[34]

These arguments should be taken seriously.

The goal behind those arguments is to break the ties between Israel and its land, and discredit the ancient ties of the Jews to the area, thus undermining the main claim of the Zionist movement. Of course, no Western person will "buy" these arguments, and if they were to be presented in the West they would be greeted with ridicule. But here too there is a trend towards "multi-messaging:" these claims are for local consumption among the Palestinians themselves, who are willing to receive such theories, and the discussion of these ideas strengthens national cohesion.

Given the above-mentioned developments in Israeli society, it is easy for Palestinian hasbara to get its messages across, particularly messages that exploit the contrast between Israeli society, which is advanced and rich, and the distress of the Palestinians in the refugee camps and at the roadblocks; and the image of the "conqueror" or "Zionist soldier". In the eyes of many Israelis, the Palestinians were now national freedom fighters.[35]

The military defeat of the Palestinians in the 1982 Lebanon war, their successful diplomatic and psychological offensive in that same war, and the rousing success of the Israeli protest movement against the war all made the Palestinians cite "strengthening the peace camp in Israel" as the most important strategic goal of their activities. On the other hand, the Jewish settlers in Israeli-occupied territories were seen as the most dangerous group for the Palestinians and the Arab world, and therefore great emphasis was placed on the campaign to delegitimize Israel.

The reason for this was not necessarily the settlers' physical existence in the territories or their effect on the Arab demographic advantage, but mainly their religious motivation and orderly ideological creed. As a result of these characteristics, they are a powerful obstacle in the Palestinian persuasion campaign. The religious settlers draw their legitimacy from the distant past, from the Bible, and in that way they are somewhat similar to the fathers of Zionism. However, the latter tried to create a new human being, and take the European Jew "back to the Bible", jettisoning all the halakhic works that had been built upon it in later ages. The "small problem" of erasing two thousand years of history did not exist in religious Zionism, which wished to return to the Bible without harming the foundations of the traditional Jewish identity. The religious Zionists are a new ideological being in the Jewish world, in the sense that they find both their legitimization and their liturgy in the Bible,[36] while preserving the halakhic development.[37]

In this sense, the ultra-Orthodox relate to their place of living only as a municipal area – without its Biblical-historical meaning – although statistically their percentage in the population living in the territories is on the rise. The ethos guiding them is still, to a large extent, the Eastern European ethos.[38]

It was in this context that the Palestinians implemented the "liability rather than asset" technique. The settlers were repeatedly presented as "obstacles to peace", and Palestinian messages also emphasized the economic expenditure involved in the settlements, which took money away from other programs, and the personal effort required from Israeli citizens, expressed in added days of reserve duty.

When we speak of hasbara activity as part of psychological warfare (PSYOP), we have to remember that sometimes we are talking of "warfare" in the simplest meaning of the word. The easier it is for persuasive messages to reach the enemy, the more they influence soldiers and their ability to fight. In principle, the army sees itself as a closed authoritative system, which conducts a close watch on the activities of enemy soldiers; but the more the army is connected to the democratic civilian system, the less it can preserve its unity of purpose and authority. Thus, after the motivation crises of the Lebanese war, similar crises appeared in the fighting during the first Intifada. For the first time, the commander of an Israeli brigade asked to be released from service for reasons of conscience, thus setting a precedent that kept the military education system busy for many years thereafter. Moreover, the army commanders, at the order of political leaders, took steps to lessen the influence of the religious component in the army. Lubavitch chassidim were forbidden to enter army camps, and the advancement of young religious officers was impeded by a glass ceiling. The pacifist play *Johnny Comes Marching Home* became a part of the educational plan for fighting soldiers in the IDF for a number of years. When the Chief Education Officer wished to remove the play from the program, a hue and cry was raised.

Against the claim of the Chief Education Officer, who thought that the play was pacifist propaganda, the spokespersons for the Israeli left claimed that it was not propaganda, but rather "a basis for dialogue and for open discussion". Needless to say, no play that had a religious, nationalistic, or militaristic side to it was ever presented as a basis for dialogue and for open discussion, and any suggestion of the sort would have been rejected immediately. In this case, the Palestinian hasbara could celebrate its victory.

The warfare side of psychological warfare has made itself felt in everything connected to Israeli POWs and soldiers missing in action: the Arab psychological warfare arm (in this case that of Hezbollah and not of the Palestinians) was able to convert a relatively minor military success of taking a few Israeli prisoners into a huge PR victory. It is doubtful whether this achievement would have been possible if the enemy had not exploited the sacred value in Israel of securing the release of prisoners. This value is as old as the Jewish people – the Mishna says, "prisoners should not be redeemed for more than their value",[39] and a significant part of the Jewish Responsa literature (*She'elot ve-Teshuvot*) deals with the topic of redeeming prisoners.[40]

This value is still upheld in the IDF today. There is an IDF unit that tries to locate missing soldiers, alive or dead, some of them dating back to the War of Independence. The Shi'ites in Lebanon understood this principle, and they are quite adroit at exploiting it in their struggle against Israel. The first time was what is called the "Jibril Deal" in 1985, in which four Israeli soldiers were exchanged for 1,150 Palestinian terrorists; according to Prime Minister Netanyahu, this deal contributed significantly to the outbreak of the 1987 Intifada.[41] The navigator Ron Arad was taken in Lebanon in 1984 and disappeared, and Israel is still expending immense resources trying to find out what happened to him. Arad is a valuable asset in the psychological warfare that Hezbollah conducts against Israel. So too are three soldiers who were kidnapped in October 2000, and a senior reserve officer who disappeared after he was enticed to go to a Gulf state. As mentioned above, these Arab military successes contribute to propaganda successes, because these events work on the Jewish-Israeli psyche to strengthen the feeling of helplessness in the face of the enemy who stands strong and never loses self-confidence. Here too it is noteworthy that the immoral and anti-humanitarian aspects of the enemy's moves in these incidents has not been exploited by Israeli hasbara, and the Israeli efforts to have the prisoners released have been focused mainly on pleading with the enemy.

## Weaknesses of Israeli Hasbara

Politics in Israel have been influenced more than a little by negative world public opinion about the country. Israeli society, and especially the Israeli state hasbara bodies (the Ministry of Foreign Affairs above all) showed especially high sensitivity to growing anti-Israel sentiment; this sensitivity was a result of the "longing for love" phenomenon mentioned above. The strong desire to look good and to achieve recognition, acceptance, and favor for the State of Israel became an important goal of Israeli foreign policy. True, every country has to be concerned about its interests and a positive image abroad is a strategic asset (both for diplomatic relations and for civilian needs such as trade, tourism, investments, knowledge exchange, etc.), but in the Jewish-Israeli case the longing has reached exaggerated, perhaps pathological proportions. From the moment this longing was noticed by the Arab world – and especially by the Palestinians – it became a powerful lever with which Israel could be subdued.

Another very Israeli phenomenon is the longing to go abroad. The ultimate prize one can offer an Israeli is still a trip abroad. This can be explained by the siege mentality caused by Israel's geopolitical isolation, and perhaps by the "wandering Jew" that is hidden in the heart of every Jew today. But in any case, this is a singular phenomenon by any standard[42]. The government uses this tendency in order to reward its workers: it seems that the percentage of Israeli government representatives who are abroad, for one reason or another, is much higher than one would expect considering the size of the country and its population.

Another matter is the double standard with which Israelis relate to America. On the one hand Israelis admire America's size, power, and wealth, but on the other hand, there is much contempt expressed towards the American personality and American behavior. Feelings of this type towards American Jews may be the result of either jealousy or disappointment that they did not leave the "golden country" and come to the Holy Land to help the Israelis in coping with Israel's difficulties. This ambivalent attitude gets in the way whenever Israel tries to explain itself (in those cases when it is decided that hasbara should be employed). The Israelis display an approach that seems paternalistic to the American Jewish community, and this has caused a rift between the two groups. It is also true that Israel has made almost no effort to satisfy the immense desire of American Jews for deep and reliable information on what is happening in Israel. American Jews, who live in a more liberal and open society than Israelis, were not interested in boring, banal, and stale messages, and were aware that the messages they were receiving did not give a full view of the internal discussion in Israeli society.

Israel's ethnocentrism and self-confidence, which were so effective in the time of the Palmach (elite military units established during World War II), became its stumbling block in its contacts with world Jewry. Israelis were viewed – rightly or wrongly – as arrogant, and when they replaced their khaki work clothes with evening dress, their basic approach did not change. Their self-confidence developed into an attitude of "we know the international arena and we'll manage". The Jewish community watched helplessly as the country's image was eroded by poorly worded pronouncements, poor English (except for Abba Eban in live appearances and Binyamin Netanyahu on television), and a lack of understanding of the language of the media. The Palestinians, in contrast, learned their lessons, used the world Palestinian community in order to develop contacts and penetrate the media, and stuck to a few simple messages that could influence the feelings of viewers without their needing any previous knowledge of the Middle East's past and present. They used talented spokespersons such as Hanan Ashrawi and Edward Said, and built an efficient Arab lobby along the lines of American Israel Public Affairs Committee (AIPAC).[43] The Arab diaspora received a simple and clear message free of debate and internal arguments, which was very different from the messages received by American Jews from Israel.

As for the non-Jewish US population, a significant percentage of the US population is simply not involved in the discussion of the problems of the Middle East. This sector knows very little about what is happening in the area, is even less interested in it, and is indifferent about the possible outcomes of the struggle. Needless to say, this sector does not have much say regarding the developments that concern Israel. On the other hand, there are groups of people who know, are interested in, and are very much involved with what is happening in Israel, but Israel is not always comfortable with their concern, namely, the Christian Evangelicals. Most Israeli leaders feel suspicious about the Christian Evangelicals because of their own distance from religion in general, because of their instinctive revulsion from the character of this religiosity, and because the American Jewish community sees them as a threatening missionary element. Above all, the hasbara people in Israel were afraid of becoming identified with the "backward" forces in American society, and continued to hope for an improvement in Israel's image among the more "respectable" elites, such as those in media and academia. The absurdity of this hope is made even clearer when one notes that it was not translated into practical action. Whereas the Palestinians had already developed a strategic hasbara plan in the 1970s using student organizations on campuses worldwide and coordinating messages and campaigns, in Israel just one official in the Foreign Ministry was appointed to deal with academic hasbara worldwide – on a part time basis. The Evangelical Christians were never seriously targeted by Israeli hasbara, and their support of Israel came not because of the hasbara policy but in spite of it. Thus it happened that now, when the "longing for love" could finally be somewhat satisfied, it was thwarted by the Israelis themselves.[44]

The Palestinians, by contrast, established a continuous presence with the Christian world in the US, among other things by using liberation theology, which had been developed, ironically enough, by the Soviets during the Cold War. A striking example is Dr. Naim Atik, a Palestinian Anglican priest of who runs an effective divestment campaign within the Christian world.[45]

Mass persuasion has already been taking place for decades on television. Television and the internet are media in which the picture is the main thing, but the materials published by the Israeli Ministry of Foreign Affairs over the past two decades are overwhelmingly verbal. The ratio between visual material and printed material published by the Ministry for the purpose of hasbara in the past fifty years is in tenths of a percent. One can also see the Jewish influence here. Visual material is unconsciously understood to be "Christian business"; Jews prefer to deal with abstract ideas and the written word, and perhaps with the spoken word as well. The fact is that in all the years of Israel's existence very few of its painters or sculptors achieved worldwide fame; the Israeli movie industry has also failed to shine. Israelis apparently still prefer the power of the word.

A watershed for the Ministry of Foreign Affairs was the lynching in Ramallah in 2000, the footage of which became available thanks to a mistake on the part of the Palestinian security forces, who neglected to confiscate it from the reporter who had filmed it. For the first time the Ministry broadcast images of Palestinian atrocities.

The pictures were very effective, because the lynching itself was not photographed, only the bodies being thrown out the window and the bloodstained hands of one of the perpetrators as he held the bodies up to the cheering crowd. Without planning it, Israel got hold of an effective clip; the murder itself would not have been shown on television all over the world because broadcasters would have been afraid of shocking their viewers. The bloodstained hands powerfully symbolized the barbarity of the murderers, and by extension the entire Intifada. But there is no proportion between the use made of this film and the use made by the Palestinians of the tape of Mohammed A-Dura, the Palestinian boy from Gaza. While the death of Mohammed A-Dura became an event that contributed to the formation of the national Palestinian identity, the lynching in Ramallah had no similar effect on the Israeli national identity. Israeli hasbara operatives let it sink slowly out of sight and made no effort to bring it back, let alone foment interest in it, and today it is slowly disappearing from the collective Israeli memory.

Following the siege on the Church of Nativity in 2002 and the successful propaganda points that the Palestinians made about the desecration of holy Christian places by Israel, the Ministry of Foreign Affairs produced a short video in which the Palestinians were shown to be desecrators of holy places. One may assume that the producers thought about the nature of their project, but it is not certain whether they understood its historical implications: for the first time, Jews directly and explicitly united themselves with Christians in attacking Muslims. Historically, the relationship between Jews and Muslims was usually good; it was the Christians who perpetrated pogroms and massacres.[46]

However, even if the message was a bit clumsy, at least this was the first serious Israeli attempt at a hasbara campaign that was of an aggressive nature, sophisticated, and aimed at the feelings and prejudices of the target audience. Such campaigns could have succeeded much earlier, but they were never seriously attempted, mainly because of the benign nature of the Israeli hasbara, influenced by the traditional Jewish and Zionist ways of thinking.

## Towards a New Approach

Beyond the traditional weakness of Jewish hasbara, what we are dealing with here in the end is a matter of Jewish identity. As mentioned above, the question of Jewish identity came to the fore after the walls of the ghettos came down, and it became even more urgent after the emergence of Zionism. Beside the question of the components of this identity, there was also the question of its "marketing". The two questions are inseparable, because the Jewish identity – as with any separate identity – is built through its differentiation from non-Jewish identities. Therefore implicit in the formation of a Jewish identity is a relationship with those who do not have this identity – the non-Jews. Obviously this aspect of Jewish identity has implications regarding what one should do about the Jew's image in the eyes of the non-Jew.

Yochanan Manor created a scale of attitudes toward hasbara: at one end there is the attitude that "hasbara can do anything", and at the other there is the feeling that hasbara is a waste of time and effort, and that only deeds can determine anything.[47] I suggest adding another scale, that of the possibility of contending with anti-Semitism. At one end of the scale there is the belief that nothing can be done about anti-Semitism ("It is a known law, that Esau hates Jacob"), and that one must continue the ancient Jewish strategy and wait for it to pass. At the other end of the scale there is the belief which is forcefully expressed by the supporters of emancipation, according to which anti-Semitism is a learned characteristic and not inherent, and the Jews should expose the beauty of Judaism to the Christian and Muslim worlds. Combining these two scales will give us a table of four combined approaches:

1. Hasbara is not effective, and anti-Semitism is unchangeable.
2. Hasbara is not effective, but anti-Semitism is changeable.
3. Hasbara is effective and anti-Semitism is changeable.
4. Hasbara is effective and anti-Semitism is unchangeable.

The first approach is that of the ultra-Orthodox. They support their position with evidence from Jewish history – rich in persecution – and from the new forms of anti-Semitism such as "anti-Semitism with no Jews" (Poland, Japan). As a group they are closed, and they see in their social and cultural entrenchment an existential ideal; therefore they do not bother to invest in improving their image. Nonetheless, they sometimes make an attempt at redress when they are worried about a legitimate complaint of non-Jews about Jews. For instance, when orthodox Jews were caught on tax evasion case much PR effort was placed on their guilty pleas and their public made regrets.[48] In this case their motivation was what is called in halacha "darkei shalom" (meaning "because of hostility"). It is in the immediate interest of the Jew to prevent danger to life. Until a few years ago, the ultra-Orthodox had not acquired any systematic knowledge regarding the media.

The second approach is that of the Israeli liberals who think that in some cases it is possible to change anti-Semitic approaches, not necessarily through hasbara, but by changing Jews and Judaism. A Judaism that welcomes universal values and encourages the integration of Jews in general society will eventually bring about the refutation of anti-Semitic opinions, and as a result anti-Semitism will disappear. This belief is integral to Israeli hasbara. The leading representative of this opinion in Israel is Shimon Peres, whose influence on the foreign relations of Israel was and is considerable. He was the one who coined the phrase "You don't need good hasbara; you need good policy". Yossi Beilin, who also implemented this philosophy during his term as Deputy Minister of Foreign Affairs, brought this approach to its highest level of development.

The third approach is the Jewish, Western-liberal assimilated approach, which sees the achievements of propaganda in their political and commercial contexts. Its advocates argue that anti-Semitism is an opinion like any other, and so is subject to change through messages in the appropriate media. Considerable resources should be devoted to changing this opinion. The reason that anti-Semitism is still prevalent – even the type that tries to hide itself under the cloak of anti-Zionism – is that the problem has not been addressed with sufficient determination, and the hasbara activity has not been good enough.

The fourth approach is that taken by the moderate right wing in Israel. The members of this group see hasbara as effective, as can be shown by many examples from a variety of disciplines, from political to commercial. However, there are areas in which hasbara is not effective, and anti-Semitism is one of them. This philosophy is reflected in a phrase cited earlier: "It is a known law that Esau hates Jacob", or on a secular level, in the belief that anti-Semitism is a cultural phenomenon so prevalent that it is almost impossible to imagine it being extirpated from the collective subconscious. Paradoxically enough, most Israelis and many of the decision-makers in Israel hold this opinion even though they are not generally considered religious. Yet we know from personal experience and from research that people are capable of holding two contradictory opinions simultaneously. The practical solution, according to this approach, is to delegitimize anti-Semites, or to cause them to feel uncomfortable both personally and socially in order to further Zionist objectives.

This philosophy is not based on empirical research, and in fact it runs counter to one of the main goals of Zionism, which is to find a solution for anti-Semitism through political activism. However, my study of the public discussion in Israel about hasbara for the past two decades has forced me to conclude that this is the real attitude that most Israelis have on the subject. They will deny it vigorously, but I conclude that in their hearts they believe that anti-Semitism is unchangeable. This view is not a product of the twentieth century, it is as old as Judaism. Despite Herzl's Zionistic attempt to find a political solution to the problem of anti-Semitism,[49] the failure of this solution – in the sense that anti-Semitism has not ceased to exist – has caused the spreading of the conclusion that any such solution is impossible. As Ben-Gurion allegedly put it: "It doesn't matter what the gentiles will say; what matters is what the Jews will do".[50]

But it seems that this belief is not only wrong, it is also damaging, for it leads to apathy and inactivity. The vast literature on anti-Semitism teaches us that anti-Semitism, like any other social thinking, is subject to change and influences, and there are many historical examples of this. An effective Israeli hasbara strategy should acknowledge that not all non-Jews are anti-Semites, and if many of them are latent anti-Semites, one can cause them to feel uncomfortable when they come to take action against Israel.

## Conclusions

Several conclusions can be drawn from the discussion in this chapter.

If Israel is interested in improving its hasbara, Israelis must decide whether they are willing to invest the necessary resources in the campaign. If hasbara can make a difference, one must invest thought, organization, and funds in it. Hasbara is a profession that requires much skill and experience: it combines disciplines such as public relations, marketing, advertising, international relations, anthropology, sociology, and psychology. In order to achieve its full effect it must be integrated in the state's strategic policy. It must be part of political, military, and diplomatic calculations, both covert and open. The government must provide training for hasbara operatives and sums of money adequate to support both one-time and ongoing hasbara campaigns. The situation today is one of reluctant, half-hearted action, unconvinced of the effectiveness of hasbara, and it is a total waste of resources.

Once the training and budget systems are set, the goals and strategy of the organization would need to be determined. The appropriate techniques for "marketing" visual and emotional messages would need to be decided, and also for multi-layering the message and conquering cultural "avant gardes", from which the message can spread to other groups.

The government must abandon the traditional benignity of the Israeli hasbara message, and be willing not only to make clear to domestic, neutral, and enemy audiences the Palestinians' failings and duplicities, but also to seek to weaken their unity and conviction in the rightness of their cause. Methods must be found for circumventing the censorship mechanisms in Palestinian society.

More than anything, this organization must conduct internal hasbara and systematically build the Israelis' positive attitude towards and confidence in themselves, their state, and their history. This goal is more comprehensive and demanding than the others, and its implementation necessitates a cultural revolution.

In summary, perfect hasbara necessitates action towards three main audiences: international public opinion, the enemy, and the internal audience. Each one of these has its own complexities, each needs a different strategy, and each demands a different "price". In the Jewish-Israeli case, the question of hasbara places before the people not only organizational and economic dilemmas, but also cultural ones. In effect it places the public in front of a mirror, where the people must face their identity, their relationship to their religion, their history – ancient and contemporary – and their relationship to the other nations of the world. In the case of Israel, the strategic choice of effective hasbara involves a very significant choice of values and ideology. The Israeli government and people could, of course, continue to avoid making the decision, and continue to conduct hasbara the way it has been conducted up till now, but the results of the Jewish and Israeli hasbara that we see today do not encourage this choice. The practical recommendations of this chapter can be summed up in a few simple phrases that are appropriate in other situations as well: study the advantages of each course of action, face the question courageously, make the necessary ideological decisions, and be willing to pay the price.

# 5

# Countering Islamic Terrorism
## The Psychological Warfare Perspective

This chapter analyzes counterterrorism measures from the perspective of psychological warfare (PSYOP) and proposes a method for analyzing terrorist acts (current and future) which would point towards appropriate and effective countermeasures.

Democracies are generally hesitant to engage in PSYOP (formerly called propaganda) since it has connotations of lies and deception, and may evoke associations with Dr. Goebbles.[1] Yet in times of war, democracies throughout the twentieth century devoted resources to this seemingly subversive field with considerable success. In its simplest form, stripped of historical and ideological burdens, PSYOP is a set of techniques for formulating messages to be sent to specific audiences through the most appropriate channel of delivery in times of conflict in order to achieve military and political objectives. Since the beginning of recorded warfare, efforts have been made to persuade the enemy not to engage in war, to despair, or simply to surrender.[2] The present form of PSYOP can be traced back to World War I, which was the first conflict to involve the total mobilization – economic, political, and military – of the nation states that were engaged in it. The active participation of citizens was essential for winning the war.[3] To this end, the nation states used propaganda on the home front (to encourage and mobilize citizens), on the battlefield (against the enemy), and in the international arena (to attract neutral parties, mainly the US).[4]

Although the details have changed, the basic techniques of persuasion developed at that time remain in use to this day. The most significant development since World War I lies in the means of delivery of messages to target audiences. Whereas during World War I the most common techniques used for delivering messages were bullhorns and leaflets, now PSYOP agencies can make use of television, loudspeakers mounted on drone airplanes, SMS messages and the internet, to mention but a few possibilities.

With regard to terminology used to describe these activities,[5] the term Information Warfare (infowar) is an all-encompassing term including deception, psychological warfare, and cyberwar (where computers are the targets or the weapon). This term is politically correct and does not have the sometimes sinister associations of other names (such as propaganda and PSYOP), but is too general for discussions of practical issues such as budgeting and manpower allocation.

PSYOP was always regarded suspiciously by the security establishment, either out of fear of its great power, or out of a belief that it was an utter waste of time. It is quite difficult to assess the effectiveness of a PSYOP campaign, and further, field officers tend to ascribe psychological effects on enemy troops to the use of "hardware" rather than to "slips of paper". This attitude has radically changed since the Gulf War of 1990–1991, as nearly 70,000 Iraqi soldiers surrendered as a result of military PSYOP efforts. The war in Afghanistan and the Second Gulf War saw a vast use of PSYOP before, during, and after the fighting; it was aimed at lowering the enemy's morale and making it give up the idea of resisting the mighty US and coalition armies. Right now, as the situations in Iraq and Afghanistan are complicated, calls for enhancing PSYOP efforts are being heard in the Pentagon and the White House. This chapter will discuss the different elements of PSYOP and their relevancy to the struggle against Islamic terrorism.

## The Foundation of all PSYOP: Research and Intelligence

Knowing the enemy, his culture, his psychological levers, tastes, and motivations, is absolutely essential for any PSYOP campaign. Much has been written on cross-cultural communication. The first methodological steps in this discipline were made as early as 1908 by the British in Nigeria and later by American government of the 1930s in its dealings with the Indians.[6] During World War I when Americans were countering different cultures, anthropologists were summoned in. Such was the case in dealing with Japanese soldiers who were rumoured never to fall prisoner to US soldiers.[7] Ladislas Farago recounts the 1930s German methodical anthropological mapping of European countries. Applied academic research beginning in the 1960s by authors such as E.T. Hall was based on the pacification work in occupied Japan which paved the way for a vast literature on the subject, which was written not only for military purposes, but also to facilitate global economical and political relocations.[8]

Hall, for instance, worked on understanding Japanese culture in the years following the occupation of Japan after World War II. Previous work had been done during the war by US Psychological Warfare units in order to persuade Japanese soldiers to surrender. These PSYOP units had to fight the common wisdom of the time, namely that Japanese soldiers would rather die for their emperor than surrender and become POWs.[9] Sometimes people are willing to die for a cause because they have been indoctrinated, and these targets require an immense investment in de-indoctrination. But in any case, through proper cultural analysis and crafting of the PSYOP message, it is possible to persuade those who are willing to die for a cause to change their minds. The solution to the Japanese problem was to lower the morale of the Japanese fighting forces and make the Japanese soldier question the validity of his training, and whether "death before dishonour" was really worthwhile.[10]

Military use has been made of cultural knowledge since the early 1930s, when the Nazis were both accumulating cultural knowledge and using it as an effective aid to their aggressive diplomatic endeavors.[11] National profiles of countries and peoples were composed in order to trace weaknesses and find possible levers.

During the Korean and Vietnam wars the American army had to adapt quickly to fight enemies with which, unlike Germany, it had no pre-existing cultural links, and of which it had little knowledge. Much scientific research (state- and business-sponsored) on the cultures of various countries has been conducted since then. US PSYOP units, which were reactivated by Ronald Reagan in the early 1980s, have mapped the potential conflict areas around the world and prepared quick deployment guides for PSYOP forces in which cultural knowledge is paramount. For the Gulf War, which was a rare case of a war that was planned in advance, the PSYOP units studied Arab culture and prepared their messages and rules of conduct for the soldiers accordingly. A few examples of some minor instructions: not to hand anything to a Muslim with one's left hand, not to show the soles of one's shoes, not to be surprised by their attitude toward women. In the Second Gulf War things became more complicated as the US and UK armies remained in the areas they occupied, and the occupation itself, combined with a lack of cultural sensitivity, caused a high level of animosity. The situation was aggravated by the absence of a plan for consolidation PSYOP, i.e. the deployment of PSYOP in an occupied area with the goal of stabilizing the situation.[12] In the US Air Force, PSYOP doctrine is now recognized as part of the Center of Gravity (COG) which the planners must consider in their preparations.[13]

## Advantages of the Application of PSYOP to Counterterrorism

Ordinarily when a government realizes that it is the target of terrorism, it should go on the offensive. Yet this must be done with care. Usually one of the purposes of terrorists is to get the government entangled in a fight against them, as this in effect grants them recognition and is an admission of their political significance. Once the government does this, the terrorists have achieved a significant victory. The government should therefore fight the terrorists without appearing to do so. One possible way to do this is by the use of clandestine units, but this is not recommended in the literature because the use of these methods can not be kept secret for long, and when it does become known, the government in question risks being seen or portrayed as ruthless or totalitarian.[14] Another better way to respond is by using PSYOP methods, which, when applied properly, can be kept subtle.[15]

PSYOP should be preferred for practical reasons as well: it is less expensive than the mobilization of combat forces, and it is a less violent response to terrorism than the alternatives. If terrorists can be persuaded to lay down their arms, and if their sympathizers can be persuaded to withdraw their support, the terrorist threat can be neutralized with less bloodshed – of terrorists, security forces, or innocent civilians.

## Target Audience through the Eyes of the Terrorists and the State

When terrorists and the government they oppose define the target audiences for their PSYOP efforts, they do so in almost mirror image terms: what is "home" for the government is "enemy" for the terrorist, and vice versa. As a result, the messages each side will deliver to the same population will differ according to the audience's relationship with the side sending the message.

## The Formulation of Messages by the Terrorist and the State

In its initial stages, a terrorist/counterterrorist conflict is less about achieving military objectives than about defining images. The messages that the terrorists would like to convey is that their cause is just, and that the government they oppose is evil and consequently illegitimate, with the aim of reducing popular support for the government increasing support for their own cause. As Tugwell has shown,[16] terrorists are likely to use specific themes when appealing to each target audience. The state, of course, will seek to refute these messages by showing the immoral nature of the terrorists.

## Semantics

One of the major tools in influencing images is semantics, and language itself is a battleground between the state and terrorists. The most obvious example is the use of the term "terrorist" versus "freedom fighter". Another example would be the use of the term "Zionist Entity" instead of "Israel" by rejectionist terrorist groups. Hezbollah fighters are called "terrorists" by Israel, but refer to themselves as *Mujahideen* – those who have undertaken the *jihad*, an honorable religious duty. The obligation to choose one or the other of these competing terms forces one to take sides in the conflict, and even act as a propagandist for one side or the other. Therefore the international media has been forced to develop strict guidelines to avoid this problem. Reuters, for example, recommended avoiding the term "terrorist" whenever possible and use instead "guerrillas", "gunmen", or "bomber". "Terrorist" should be used only if it is in a quotation from a source.[17] Another example is the struggle over the name of the joint US, UK, and Afghan operation in Afghanistan. It was initially called "Infinite Justice" but changed to "Enduring Freedom" since the Muslims associate infinite justice solely with Allah.[18]

## Means of Delivery for the Terrorist and State

As mentioned earlier, the mass media are the primary channel of delivery of PSYOP messages. Here the government has the upper hand, since it has better access to state-owned and private media outlets. It may also choose to use various levels of censorship.

Terrorists compete with the government's advantage by developing creative methods to gain the media's attention. The terrorist organizations have the upper hand in initiating stories, since they are on the offensive; they have the advantage of flexibility and speed of response over the government due to their small size, and sometimes they receive the sympathy accorded underdogs. Both the government and terrorists will attempt to manipulate the media by exploiting its weak points: pressing deadlines, journalistic ignorance of military matters, a preference for dramatic images, and a tendency to identify with the underdog.

In the early stages of the struggle the terrorists in the past were likely to have access only to alternative media such as leaflets, posters, graffiti, symbols (such as the color of a forbidden flag), and so on. The Information Age revolution has changed the PSYOP landscape with e-mail, USB flash drives, and cell phones. At later stages, the terrorists will frequently utilize sympathetic neighboring countries as a base for distributing their own electronic media. Throughout they will employ recognized media management techniques to create links with the media. They will initiate human-interest stories, and set up image-generating events. The massive coverage of the kidnapping and execution of journalist Daniel Pearle in Pakistan and the kidnap of Israeli soldier Gilad Shalit are examples of such use.

All this of course applies in a democracy, where the media is free to report everything, except perhaps most sensitive security matters. Furthermore, the technological innovations of the past two decades have made censorship and the restriction of information access very difficult. Democratic governments therefore have to develop strategies of media management to deal with the terrorists' PSYOP strategies, instead of relying on censorship or the patriotic feelings of journalists.

## Implementation of PSYOP for Counterterrorism: The US vs. Bin Laden and Al Qaeda

The name Al Qaeda (the Base) reflects the scope of the threat that this terrorist organization poses to the West. It maintains an ideology broad enough to include various sects of Islam in various regions and countries, and it is organized in loosely connected cells which enables fast proliferation and at the same time keeps the other cells safe if one is compromised or destroyed. Bin Laden has used his personal wealth and technical skills acquired in Afghanistan to establish a group that may spread throughout Muslim communities worldwide and destabilize current world order. There is no guarantee that Al Qaeda or one of its affiliates will not assemble a larger coalition of "oppressed" Muslims from various other denominations or social groups. Therefore defeating Al Qaeda is of prime importance not only to the US but also to the West in general.

This struggle demands new strategies. Conventional warfare is largely irrelevant, except in specific situations such as the deployment of Special Forces in places like Afghanistan for counterinsurgency missions using tactics developed since the 1950s. Counterterrorist strategies would seem to offer some hope, but this case is made more difficult than any previously addressed by the addition of religious fanaticism, suicidal techniques, and global dispersion. Even classical PSYOP is powerless in confronting suicidal terrorists, as the practical knowledge and techniques in the field of PSYOP developed during the last hundred years were based on the basic human will to survive. Thus Al Qaeda constitutes a challenge hitherto unknown to Western society.

Nonetheless, in combination with standard measures such as intelligence gathering, Special Forces deployment and diplomatic negotiations, PSYOP can be effective, though its deployment will require the government to undergo mental and organizational changes.

### Preparations

In order for a PSYOP campaign to be waged effectively, it must begin before the hostilities – in the case of terrorism, before the terrorist threat manifests itself. Before the hostilities begin, preparations can be made without pressure or complications from the government bureaucracy or from public opinion. In the case of Al Qaeda it is too late, but this is worth bearing in mind in relation to future offshoots the organization is planning. The US government should prepare a policy on PSYOP and terrorism and set up a single agency to address all aspects of PSYOP that relate to terrorist activity. This agency should be included in the highest levels of policy-making, and should include a fast-response spokesman. The familiar "war game" methodology should be expanded to include PSYOP scenarios, and the PSYOP agency together with other government agencies should run simulations of terrorist attacks and of the messages that are most likely to be exchanged – and evaluate their effectiveness. The agency should educate the media[19] regarding the ways the terrorists will try to use them in what Jenkins termed the theater of terrorism.[20] Though the media is generally familiar with the ways terrorists manipulate it, and participates with mixed feelings, the government should stress to the media and to the general public this aspect of the conflict in advance in order to teach them that the gory images that are often disseminated following a terror act serve the terrorists more than they serve the public interest. This process would require briefings both by academics and intelligence officers (see the section below titled "Research and information"). Getting the cooperation of traditional intelligence organizations will likely be extremely difficult: from the first day on the job their members are taught the importance of secrecy. Yet, the challenge posed by Al Qaeda compels these organizations to change: failing to do so might cost dearly. The US intelligence organization must adapt to the new situation. They should cooperate with the PSYOP agency that would use them to brief the media without jeopardizing their sources. Such public instruction regarding the techniques and purposes of the terrorists might reduce the terrorists' success.

Learning the lessons of the First Gulf War regarding the importance of visuals, the agency should continuously prepare a large quantity of broadcast-quality video together with 'youtube' compatible clips that reflect the position of the government, preferably in an implicit manner. Such material may document police drills, the preparations of medical organizations, and so on, which would provide a sense of government preparedness for future attacks. If the aim is to damage the terrorists' reputation in order to align the public behind the government, then appropriate material should be made handy, such as videos and other documents relating to Al Qaeda's experiments in biological warfare.[21]

An incident from the history of PSYOP illustrates an important principle, that of exploiting opportunities. The British publication of the *Lusitania* medal in 1915, which had been produced by a private German mint to commemorate the successful sinking of an ocean liner sailing between England and the US, played an important role of swaying American public opinion towards participation in World War I. The US PSYOP agency should monitor and analyze media stories and intelligence reports twenty-four hours a day. The staff, trained in PSYOP skills, would identify which stories or reports had potential PSYOP value, and would have the authority to act accordingly. The tremendous flow of information from the internet (including blogs and news sites) and twenty-four hour television news, when combined with the desire for a rapid response, entails risks, as there is little time to check facts, but not to respond quickly is to loose the battle, and quick response has the potential advantage of keeping the enemy on the defensive.

### Research and Information – Cultural Knowledge

Another area in which special attention should be given is research and information. Of course, every state has conventional intelligence agencies, whose job it is to collect information. However, PSYOP requires a different sort of intelligence from conventional warfare. Whereas the latter seeks to answer the questions when, how much, and where, PSYOP intelligence is primarily psychological and aims at determining how the target audience's psyche works. It uses psychological analysis (which is also used by conventional intelligence in attempting to guess the opponent's moves) to find out how and to what extent the target audience's culture influences its psyche – and how messages can influence its psyche. The more different the opponent's culture, the more difficult the analysis. During the two world wars it was fairly simple for the warring parties to answer these questions, since both sides in the European theatre belonged to a culture based on the Judeo-Christian culture. Things were much more difficult from the PSYOP perspective when the US fought in Korea and Vietnam.[22] Once the psychological, cultural and anthropological data has been gathered the PSYOP agency can search for the special needs bit of information that will help it craft effective messages and thus contribute to success in the war.

The government should seek to facilitate discussions between PSYOP operators and intelligence officers in order to use the information to maximum effect. Thorough research can produce in-depth information that can be used in personal slander campaigns against terrorist leaders; it might also enable the government to expose the involvement of foreign elements of which the people already have a negative opinion, or the existence of front organizations that the public is unaware of.

## Arab Islamic Culture and the Western World

In the case of the Islamic terrorist challenge, it is surprising that despite America's dependence on oil from the Middle East, their ongoing dialogue with the Arab oil aristocracy, and the enormous amount of petro-dollars channeled into academic study of the Middle East, America has shown profound ignorance of Arab-Islamic culture. For example, Americans were deeply surprised by the phenomenon of suicide attacks. A look at the vast media output that followed the September 11 attacks, which tried to give some explanation regarding the motives of the hijackers, reveals that the various commentators committed the basic error of trying to decipher another culture through the perspective of their own. It is very difficult to explain to a member of a Western, industrialized society the common Arab *Weltanschauung* of *Mu'amara* (conspiracy),[23] or the importance of "saving face", to the extent that inter-family murder obligations are still taken seriously in large parts of the Arab world. As to the suicide attacks on September 11, some Americans resorted to explanations such as mental derangement, deception by the terrorists' leaders, personal problems, intoxication, or unknown brainwashing techniques. Even in Israel, where suicide bombings have been occurring since the 1980s (first in Lebanon, and later in Israel proper), the impression is that there is no comprehensive answer. The public debate as to whether suicide attacks are the result of desperation caused by economic distress, emotional and sexual exploitation, or manipulation of the religious fervour of the attackers – or a combination of all of these factors – is still raging.

In the long run, America and the EU should perhaps consider the model Britain used in the eighteenth and nineteenth century of government-aided encouragement for families to live abroad and educate their children locally. Such a program could help the forthcoming generation to serve as a cultural bridge between the Arab world and the West much better than any academically trained officer. Yet, academia can assist this effort as well, and the original British School of Oriental and African Studies (SOAS), originally founded to train British administrators for overseas postings, is a paradigm worth emulating.

All in all, a much greater research effort is called for. The first steps are currently being taken in the direction of a re-assessment of Middle Eastern academic studies in the past three decades.[24] Another step now being taken is the government recruitment of Arabic-speaking Americans as translators and analysts.[25]

Undoubtedly, a considerable part of the resources should go to academic studies of Islamic societies. The energy and money devoted during the decades of the Cold War to Sovietology should now be directed towards "Islamology". Applied research in this field could contribute considerably towards diminishing the Islamic threat. The Western world should engage in a large-scale cultural research of the contemporary Islamic world, free from political constraints and from the need to remain within the boundaries of what is considered politically correct. The focus should be on the cultural-anthropological perspective. This term is vague enough to include many sub-fields such as history, art, society etc., but the basic aim should be that of understanding the Islamic world. Followers of Edward Said will naturally protest and label this effort a "tool of Western cultural domination", but with proper marketing and the current political climate, that should not pose a serious obstacle. A possible answer to such an allegation could be that both sides would benefit if the Occident finally achieved a proper understanding of the Orient, rather than a romantic Lawrence-of-Arabia image, or for that matter one dominated by images of terrorists.[26]

## Formulation of Themes and Messages

Before we discuss the formulation of PSYOP themes, it must be emphasized that PSYOP messages alone, no matter how clever and inventive, will never give a complete and perfect solution in the long run unless they are based on an ideology that can bridge the gap between the Islamic world and the West. This is of paramount importance. Without such an ideological framework, the relationship between the undeveloped Muslim world and the developed Christian West will continue to be one of slave–master relations, which in turn will breed more hatred. Therefore an ideological framework should be developed that will determine the attitudes of each respective religion, and the basis for a productive relationship between. Its principles should then be translated to marketable themes and than delivered to the respective pre-defined target audiences.

Towards the end of establishing this ideological framework, a think-tank should be established to house under one roof academics and clergymen, aided by PR consultants with experience working in the Arab world, who will inject their opinions on how to formulate the principles in a way that Muslims will not find condescending or patronizing. The goal of the think-tank, all members (including Muslims) of which should be wholehearted supporters of the peaceful coexistence of Islam and Christianity, would be that of finding the common denominators between the two religions and cultures.

Since this ideological framework is bound to be attacked by Islamists as another attempt at "Western domination", suitable preparatory steps should be taken in order to guarantee its success. The think-tank should be presented with preliminary research that details the themes, arguments, and accusations the Islamists have been disseminating during the past two decades, so that appropriate responses can be studied. Themes such as the common belief in God and the sanctity of life can be developed in the hope of putting an end to the legacy of hostility between the religions.

Based on the research suggested above, the proper themes and messages will be composed. In the following paragraphs I present some general thoughts about possible themes for PSYOP messages US should consider sending, divided according to their target audiences.

## Messages for the Home Audience

In addressing its home audience the US government might use "conscription themes". These are themes by which the government requests the people's political support for anti-terror legislation, allocation of resources towards anti-terrorism efforts, and temporary postponement of certain peacetime civil rights. Relevant themes include "unity", "emergency", "homeland", etc.

But there is also the problem of the Muslims living in Western countries to consider. The US government and its European allies will have to decide what approach to take to these groups, and it will not be a simple decision to make. Many European countries including Britain have a considerable Arab minority, which will have to choose between supporting radical Islam and supporting the government. The situation where radical Islamists living in Western democracies conscript volunteers and raise funds for movements such as Al Muhajirun and Al Qaeda can no longer be tolerated. On the other hand, Western governments will have to take these domestic Muslim communities, which are part of their home audience, into account and determine how far they are willing to go in terms of messages that might offend these communities. It will not be easy to draw the line between appeals against terrorism and free speech.

The US government should keep in mind that after a terror attack, when a few months pass and the initial shock lessens, the pro-Arab and pro-Muslim grassroots organizations and political action committees, together with an impressive array of front organizations, will be set in motion to urge a return to basic American values, human rights, racial tolerance, and religious freedom. These would hamper American efforts to uphold feelings of outrage regarding the attack, feelings necessary for popular support of the war effort.

As the second year after the attack unfolded, the US government line was to attempt to distinguish between "Old Islam", which is peaceful, and radical fundamentalist "New Islam". Though this effort is quite understandable, it raises a serious complication for the government. The distinction is of questionable validity, and in any case it is not clear whether the general public would be able to make such a distinction. If any further terrorist activities occur, public opinion in the West will tend to be against Islam of any variety. And this extreme reaction is to be guarded against.

The general image of large crowds of worshippers bowing on carpets evokes feelings of a distant, primitive, inexplicable religion. The Muslim community still has a long way to go to change earlier images of greedy, sybaritic sheiks, originating in the early 1970s.[27] The image of Muslims was also damaged by scenes of stonings of "adulterous" women in Afghanistan and of hangings of homosexuals in Iran. Some of these images were disseminated as part of Western PSYOP campaigns to mobilize support for continuing the war in Afghanistan and to mount a campaign against Iranian nuclearization. When the government launches PSYOP campaigns against Islamists, there will probably be substantial spillover towards traditional Islam, and this factor should be taken into account. In order to avoid an undesired negative image of anything Muslim, the US government together with its Arab allies should engage in a long-term campaign to educate its citizens about Muslim culture. A belated investigation of passages about Islam in US school textbooks revealed extensive misinformation, which did not occur by mere chance.[28]

The proper approach to be taken by the American government would be to re-focus on a personal manhunt after Bin Laden and his accomplices, as well as pursuit of anti-democratic ideologies and institutions. Classified material on institutions and personalities can be released, exposing their double standards. The government should not wait for private organizations to expose front organizations such as CAIR (Council on American-Islamic Relations), which was supported by President Bush after 9/11, or the Holy Land Foundation, which was a major funding channel for terrorism.[29] Focusing on terrorist leaders such as Bin Laden would enable the US government to channel some public rage, while at the same time not demonizing Islam and the entire Arab community.[30]

## Enemy Target Audience

I now present some examples of messages that the US might wish to transmit to its enemy audience:

First, that democracies react slowly but decisively. The purpose of this theme is to instill fear in the target audience. The US government needs to overcome a very powerful image of American incompetence that the Islamists have been instilling in their home audience for the past two decades. This incompetence was demonstrated by military failures such as Vietnam and Somalia and was presented by Islamists as part of an historic determinism, very much like that promulgated by communism in its early stages, which would lead to the downfall of America. The drug situation and the sexual promiscuity in the US were also presented as telling signs. The message should be that democracy is not a synonym for weakness; it takes more time to reach a consensus and take action in a democracy, but there is no defense against its physical and moral might.

Second, resoluteness: perseverance is not a uniquely Islamic trait. It is a big mistake to underestimate the tenacity of the American people. The pursuit of the perpetrators of acts of terror against the US will continue until the perpetrators are punished. The families of the thousands killed at the World Trade Center will make sure that the issue will remain on the agenda for decades to come.

Third, moral superiority: the killing of thousands of innocent people shows the heinous nature of Al Qaeda and their supporters. The victims were hard-working people supporting themselves and their families. Whoever murdered them committed a crime against God and humanity. Such an act, it must be emphasized, is forbidden by Islam. This last theme might be developed into a category of "attack themes" which could include demonization of Bin Laden and his accomplices, presenting them as morally corrupt: not only do they lead a life of luxury, but they disregard the lives even of their own people.[31]

And fourth, countering the underdog theme. Thought should be given to ways of denying the terrorists the image of the underdog. The problem here is that the government is always perceived as the strong guy. One way to solve the problem would be to focus on the individual rather than on the state, that is, to focus on the fate of some of the individuals who perished in a terror attack.

## Neutral Audience

For neutral audiences an important message that the US might want to convey is that of a "common danger", designed to create a sense of urgency among fellow Christian countries that they either "hang together or hang separately". Al Qaeda could be presented as a forerunner of an Islamic revolution in Western Europe. A recent example that would help crystallize the message is the Nigerian death sentence to the woman who gave birth out of wedlock.[32] Other examples include the plight of Salman Rushdie (free speech) and the Sudanese slave trade.

## Semantics

The names assigned to the two sides in a conflict have always been an important issue. Governments have tried to instill the notion that terrorists are not guerrillas (urban or rural), but criminals acting for profit or power, for example. The case of Al Qaeda and its affiliates is more complicated. The 9/11 attackers did not act for money or material gain but rather out of religious conviction. The old molds are no longer valid. As mentioned above, the Western mind, which was largely unacquainted with Islamist radicalism, has sought to explain it in familiar Western terms, and therefore has labeled the suicide pilots as lunatics or brainwashed. In the current political atmosphere the media are cooperating almost completely with the American government and do not grant the perpetrators of the WTC attacks the positive status of "guerillas". When the shock of the attacks eventually dissipates, the terrorists will seek to be called by the respected title of *Mujahideen*. The government should contest this title worldwide and not merely in the US, and label them as deranged terrorists, and a danger to any society.

## Images

Finally, there is the matter of translating PSYOP concepts and themes into concrete messages. This is the task of PSYOP officers or culturally sensitive civilian PR consultants.

Images are of prime importance. Through images one can effectively convey the cruel nature of the war that the terrorists have launched, for instance. The government should do its utmost to give the media access to the scenes of terror attacks and assist them in relaying images and information to broadcast centers. This immediate access to the scenes of attacks is, from the PSYOP perspective, no less important than the intelligence that might be gathered at the scene of the crime, which could lead to the capture of the perpetrators.

The government should develop a visual vocabulary of the messages it wishes to convey. The most effective image of the Gulf War – the cormorant covered with crude oil – was very helpful when it came to vilifying Saddam Hussein. In America, the turban and beard can create a sense of estrangement. Shortly after the 9/11 attack the

US media picked firemen as a tool for giving context to the story of the attack. They were the (visual) incorporation of devotion, defiance, and patriotism for the public. These and other themes should be translated into visuals by teams consisting of PR consultants, senior PSYOP officers, film directors, and news editors. The messages based on such themes should be analyzed by senior policymakers in terms of their possible political repercussions.

The visual manifestation of "attack themes", such as the moral deficiencies of the enemy, could be images of Rolls Royce cars, excessive palaces, gold bullion or casinos. These images convey a sense of corruption and they could be linked with enemy leadership when there is intelligence to that effect. Western technical superiority can be presented through images of satellite dishes and antennas such as those at Fort Mead which will demonstrate the level of US monitoring of enemy messages. Images relating to forensic science, such as images of DNA analysis (microscopes, white robes) or thermal imaging could convey the idea that "you can run but you cannot hide".[33] The theme of slow but sure democratic reaction could be conveyed by images of an aircraft carrier: its size enables it to absorb a lot; it takes much time to turn, but once it turns, it is unstoppable, and its retaliatory force is enormous. The theme of resoluteness can be represented in an image of a mother who lost her child in a terror attack or a brother who joined the army: their facial features could communicate the theme without a single word. But these suggested visual themes do not stand on their own. The government should encourage the media to incorporate these images in their reports, so that for example a report about debate on defense issues will include images from a tour in a laboratory, of weapons experimentation, or of the helicopter of an important official landing on an aircraft carrier, with the video taken so as to show the impressive bulk of the ship.

## Means and Channels of Delivery

Since the Islamists cannot approach their enemy audience directly, they will resort to the use of media management and the internet.

Bin Laden has cleverly used CNN to deliver his messages by making his tapes a high value news story in and of themselves, thus achieving far more publicity than if he were to arrange to have the tape delivered to the US government, for example. The government tried to limit the tapes' effectiveness by resorting to unconvincing patriotic arguments that the tapes might contain secret messages, and thus should not be aired. The network compromised and when the next tapes arrived they aired only the parts that did not directly promote Bin Laden's ideas. The second channel open to the Islamists is the internet, and there their message is delivered freely. Still, it is limited to those who have access to the internet and who seek out their websites. Al Qaeda has very few options now, as American law enforcement agencies are closely monitoring the global communication system for any sign of their presence.

The US government has the upper hand in terms of delivery, and has extensive contacts with the media. In times of national crisis, for example following a vicious terror attack, the generally critical media stand almost unanimously behind the government. For Arabic-speaking audiences (enemy and neutral), the government has the prestigious worldwide service, Voice of America. In Afghanistan, the US Army PSYOP units are communicating both with enemies and with the Afghani populace through the classic method of leaflets[34] dropped from airplanes or handed out in villages. In Vietnam the US Army dropped portable transistor radios to the villagers so they could listen to American stations. It is no science fiction to provide cheap netbook computers and their likes that would connect internet surfers to websites and blogs in Arabic or Pashtu originated from US content providers. Admittedly surfers could go to Jihadi websites as well; however, PSYOP history is replete with marketing techniques designed to attract audiences to contents composed by their enemies.[35]

## Conclusion

PSYOP is recommended as another tool in the arsenal of the US and Western governments in their war against radical Islam, to counter the ideological and physical attack by radical Islamists. In order for this effort to be successful, the governments should increase their PSYOP capabilities in terms of research into Islamic culture, develop an ideological framework, define the proper messages, and find a way to translate them in the cultural language of the designated target audience. Visual images are very important. Creativity and considerable resources should be invested into delivery of these messages.

Finally, it should be remembered that PSYOP, as effective as it can be, is only one tool, and needs to be used in conjunction with diplomatic and military action in order to achieve its full effect against radical Islam.

# 6

# Cultural Warfare
## Secularization Defense Initiative

BENJAMIN BROWN AND RON SCHLEIFER

In this chapter our aim is to present and discuss the concept of cultural warfare, a concept we developed which is based on aspects of information warfare and more specifically of psychological warfare[1]. In psychological warfare a distinction has generally been made between three levels of deployment: tactical, operational and strategic. The former, limited in aim and scope, is designed to affect specific military operations on the ground. The latter, extending beyond the battlefield, targets enemy public opinion, among other things, and aims to reshape that public's military and political perception of the conflict as a whole. The ambition and sweep of cultural warfare are wider still: to transform, wholly or partially, one's adversary's cultural landscape; hence, the label "meta-strategic psychological operations". Cultural warfare did not spring into being out of nowhere, but is the product of a systematic study of several post-1945 wars and conflicts, all of which involved democratic states and coalitions, including the West's current battle with radical Islam.

Though it might be considered to belong under the umbrella of Huntington's "Clash of Civilizations" theory – a theory to which we, in principle, subscribe – the doctrine of cultural warfare is neither derived from nor dependent upon it. Yet, even among those who, rejecting Huntington's thesis, regard the conflict between the West and radical Islam in less cataclysmic terms, there is a growing realization that the West is facing a new kind of war, one that requires new, innovative strategies, tactics, and weapons. Nowhere is this more true than in the field of psychological operations (henceforth, PSYOP), where the whole question of influencing enemy consciousness is in urgent need of a new, original, more comprehensive, even revolutionary approach, both in terms of aims and technique. The Bush–Sharansky doctrine, in insisting that the key to victory lay in regime change – that is, in democratizing the Muslim world – grasped this point in part. Yet attempts at democratization of Muslim nations have so far met with questionable success. Not that this is entirely surprising, as the Bush–Sharansky doctrine is flawed, patchy, not well thought out, and evidently incapable of delivering the goods. What is needed, and what we propose, is a bolder, more challenging plan that seeks to thoroughly remodel Muslim consciousness, for it is only then that democracy will be able to take root in the Islamic world. Put simply, the Muslim world must pass through a modern-secular revolution, similar to the one in the West in years gone by: a revolution that goes beyond the introduction of economic and technological innovations and reaches into the very heart of the individuals' and communities' *Weltanschauung*. However, unlike Western society, the Muslim world appears unable or unwilling to embark upon such a radical change on its own. It clearly needs to be helped: inspired, propelled, provoked, and, dare we say, maneuvered into embracing a modern-secular mindset, which is where cultural warfare comes into play.

For all its virtues, democracy also has its fair share of weaknesses. The question, for example, of whether when locked in battle with undemocratic societies, liberal democracy is a source of strength, weakness, or possibly neither is one that has long challenged academics, politicians, military personnel, and pundits alike, whose answers have varied from yes, to no, to all points in between. It is no part of this chapter's remit to delve into this highly contentious and complex issue; suffice it to say that we are firmly on the side of those who regard democracy in wartime as problematical and even possibly a liability. This contention will be taken as a premise throughout this chapter.

Democracy, defined as the rule of the people by the people, created a culture that treasured, indeed sanctified, the individual together with his or her wants, needs, and liberties above all else. This can be a serious handicap when a longtime democracy becomes embroiled in lengthy military conflicts with single-minded, ideologically driven, and ruthless adversaries. Self-absorbed and self-centered, people born into a liberal-democratic culture might sometimes have less of the required patience, fortitude, or resilience to pursue such conflicts to their bitter end. Which is why we suggest exporting that culture to the Islamic world. Put simply, democracy is the ideal tool to weaken and destroy radical Islam's ability to wage war on the West by destroying its appetite for endless conflict.

Conscious that our view of democracy is tendentious, to say the least, and likely to ruffle a few feathers, we should make it clear that we have no desire or intention to argue in favor of changing the face of Western liberal democracy, let alone of abandoning it for another form of government. Quite the contrary, our sole purpose is to suggest ways of winning the war against radical Islam, and this by putting forward a strategy that will prompt the Muslim world to embrace liberal-secular democracy as it stands, with all of its faults, thus exploiting not only democracy's virtues, but also its failings.

## The Problem of Weakness

It is self-evident, and is indeed one of cultural warfare's theoretical underpinnings, that a PSYOP offensive must be designed to support to the aims, scope, and nature of the campaign as a whole. Any change in the campaign requires a commensurate change in the cultural warfare operation. To date, most wars have been fought between states, state-coalitions, and political-cum-military movements, or some variant thereof. In all cases, the aim has been to win a decisive military and political victory. Radical Islam, however, in joining battle against the West, appears not only to have rewritten the rules of engagement, but to have moved the goal posts to boot. It freely acknowledges that its goal is global domination. A minority of the more violently extremist Islamists believe that this ambition can be achieved solely by military or political means, but even they concede that ultimately, victory lies somewhere in the distant future. The majority hope to realize their dream by undermining and weakening the West morally and culturally, a long-term project spanning decades if not centuries. Their immediate and shared goal is to wear the West down, to exhaust its energies and shake its confidence, putting an end to what they see as its crusader-like zeal to conquer Muslim lands. In the long term, the cumulative effect of these victories, plus the constant erosion of the West's moral and cultural resilience, will, they are convinced, lead to the total collapse of Western civilization and the beginning of a new era of Islamic worldwide supremacy.[2] Terrorism, the Islamists' main weapon, whose chief effect is psychological – sapping both enemy morale and its will to fight – rather than military, is plainly tailor-made for this kind of war.

Recognizing that it is engaged in a new, singular kind of conflict, the West formulated several battle plans that can roughly be divided into two categories: defensive and offensive. A typical example of the defensive approach is Huntington's "Clash of Civilizations" theory.[3] Huntington argues that the West should abandon its bid for global ascendancy, look inward, and concentrate upon shoring up its basic values within its own communities. What Huntington is proposing, in effect, is that the West turn its back on the wider world and hunker down behind its natural borders. The offense-oriented school of thought is best represented by the Bush–Sharansky doctrine. Democracies, President Bush emphasized, are peaceful, which is why the West has a duty to spread democracy throughout the world.[4] It is a policy, Natan Sharansky added, which has succeeded in the past – witness the collapse of the formerly powerful Soviet Union – and there is no reason to suppose that it will not do so again in the future, in the battle against radical Islam.[5]

Democracies, President Clinton asserted from the other side of the US political divide, do not attack each other.[6] The "democratic peace theory," as it is better known, has been the subject of intense dispute in recent decades, but it is clear that if we consider it in the light of a strict interpretation of democracy and a strict interpretation of war, we may affirm that at least in the twentieth century, the democracies that went to war against each other were very few.[7] Yet this does not necessarily mean, as some were quick to conclude, that democracies are inherently peaceful.[8] A statistical correlation between two variables will often produce a plausible hypothesis regarding their relationship, but without a convincing causal link explaining the connection between the two, it will remain precisely that: a hypothesis, not a fact. So the question arises: If they are not inherently peaceful, why are democracies so clearly reluctant to go to war?

A child of the Enlightenment, liberal democracy embraced many of its progenitor's principles and values. It does not necessarily follow, however, that peace, which the Age of Enlightenment undoubtedly espoused, was one of those values. The principal aim of liberal democracy is to protect the rights of the individual against the state, and this by various means, including elections that ensure the transfer of power in a just and equitable manner. In principle, this has little to do with questions of foreign policy, war, or peace.

Hence, it is quite possible that democracies – well aware that they are at a disadvantage when it comes to doing battle – tend to avoid wars not because they are intrinsically peaceful, but because they are understandably reluctant to put their people's strength of mind and stamina to the test.⁹

If democracies are weak, and yet liberal democracy is still considered the best of all possible political systems, the obvious conclusion would seem to be that when battling radical Islam, the West ought to take advantage not only of democracy's strengths but also of its weaknesses. Put simply, if the West's weaknesses when waging war are rooted in its democratic, liberal, and secular values, then one way to win the war against radical Islam is to export democracy to the Muslim world. By inflicting its own inherent failings on the enemy, the West will create a more level strategic playing field in one very important area, one where it is at the moment at a distinct disadvantage: public vitality, morale, and endurance. Coincidentally, exporting democracy to the Muslim East is precisely the aim of the Bush–Sharansky doctrine, although Bush and Sharansky came to appreciate the value of that aim by a different process of reasoning.

## The Road to Democracy: Modernization and Secularization

As noted, we have no argument with the Bush–Sharansky goal of introducing democracy to the Islamic East, only with its theoretical basis for that goal. Democracy should be exported to the Muslim world not because it is fundamentally peaceful, but because it makes for rather selfish, precious individuals who lack the strength of character to engage in lengthy and costly wars. Accordingly, the democratic West could do no better than to infect its adversaries with its own weaknesses. The question is how it should be done.

The answer? One need look no further than Montesquieu, who already in the eighteenth century noted that "the government most conformable to nature is that which best agrees with the humor and disposition of the people in whose favor it is established".¹⁰ In other words: political regimes do not exist in a vacuum, but evolve within the context of a particular and apt cultural framework. Bestowing the gift of democracy on a people for whom the modern-secular culture is largely *terra incognita* and who are thus not conversant with the values underpinning democracy, is akin to giving children the keys to the family business and leaving them to it. This is more or less what the Bush–Sharansky doctrine did, with predictable results. It assumed that all that was needed was to install an ample number of democratic institutions in the Muslim world and the rest would follow naturally; that, mesmerized by the virtues of freedom and liberty, Muslims everywhere would rush to embrace democracy and all it stands for. Of course, no such thing happened. Instead, Western attempts to bring democracy to the Muslim world were denounced, particularly in the Arab Middle East, as imperialism. Traduced as a viral and alien import, democracy was condemned as a device by which the ex-colonial powers cunningly sought to resume control of Muslim lands.[11]

If democracy and democratic institutions are to take root in the Muslim world, then, pace Montesquieu, Muslim society must first undergo a cultural revolution that would result in its embracing modernization and secularization. Orchestrating such a revolution is an ambitious, complex and subtle project that demands more than the wielding of behind-the-scenes political influence. If the aim is to transform the Muslim community's basic and deep-rooted values and traditions, then what is needed is cultural influence, both covert and overt, a task for which cultural warfare is eminently well suited.

Traditional PSYOP doctrine focuses on either undermining the enemy's fighting spirit or reshaping its political consciousness, within the context and time frame of a particular military or ideological conflict. None of this is of any relevance in the war against radical Islam. As noted, radical Islamists think in terms of centuries rather than years, thus posing a great challenge to any adversary that may threaten their fortitude on the direct, conscious ideological level; after all, they do not expect to see the victory during their lifetimes. Moreover, terrorism, Islam's principal weapon, is typically used sporadically rather than systematically over lengthy periods of sustained fighting. Nor is it much use targeting the Islamists' political ideology, as Islam – a social, political, psychological, and religious-cultural phenomenon – encompasses more than mere ideology. Hence the need to challenge, weaken, and ultimately demolish its entire cultural edifice, replacing it with a new modern-secular construct. If the aim is to transform the Muslim community's cultural landscape, then one must move beyond the limits of traditional PSYOP and into the realm of cultural warfare.

It is worth noting at this point that the Muslim world seriously challenges the Marxist thesis that democracy is a necessary by-product of capitalist society, with both Saudi Arabia and the Gulf States plainly disproving this sweeping assertion. Far from possessing feudal economies, as is often claimed, these countries are more akin to modern capitalist societies. Admittedly their financial assets are almost entirely concentrated in the hands of a few wealthy families. But this narrow elite, embracing the principles of the modern free market, has also welcomed some utilitarian aspects of modern Western living, indeed to an excessive degree; hence its determination to possess the very best that Western technology has to offer, and it members' ultra-opulent, ultra-consumerist lifestyles. However, and in stark contrast, these same elites are die-hard reactionaries faithful to their notions of tribal hierarchy, and opposed to social, political, or cultural modernization. As a result, their countries have remained locked in a time-warp with respect of their social and political order, the status of women, their moral code, and, above all, their religious traditions.

As the Arabian example proves, contrary to Marxist theory, material circumstances are in themselves insufficient to determine a society's social, political, or cultural trajectory. However, people's mindset does have a profound and lasting effect on and is key to the way in which communities and states evolve. Nevertheless, it seems reasonable to assume that it would be far easier to engineer a cultural revolution in societies that have adopted some of the material accouterments of modern Western society than in those that have failed to do so.

What remains now is the practical question of how to reshape Muslim consciousness? How to transform – modernize and secularize – Muslim society? One way of answering this question is to look to precedents and examine these from a general, as well as PSYOP, perspective. Accordingly, employing an empirical, historically based methodology, we will scrutinize previous efforts to refashion public consciousness and apply the resulting lessons, with the necessary adjustments, to the current task of transforming Muslim society; this approach will also have the added bonus of offering a better understanding of cultural warfare's historical roots.

It is impossible to examine all relevant precedents, so we shall focus on the two most familiar and apposite cases: the modernization and secularization of Christian society in Western Europe, and of the Jewish communities in both Western and Eastern Europe. Historians and sociologists have rightly described the processes of modernization in these societies as a cumulative series of events and developments sparked and sustained by a natural internal dynamic.[12] While this is undoubtedly true, this is not the whole picture, as the proponents and promoters of modernism exercised, in addition and at the same time, an on-going and deliberate manipulation of individual and public consciousness. Unlike a society's internal dynamic, that comes and goes without human deliberation, such manipulations can, in their various guises, be used time and again. We propose to isolate, analyze, and integrate these into an overall cultural warfare battle plan against radical Islam.

The processes of modernization in Western Europe can be roughly divided into two phases: an incubation period followed by an eruption lasting decades, if not more. It was during the initial incubation period that various aspects of modernity emerged, though without any guarantee that these would either survive or be followed by others. Nor was there any assurance that the process as a whole would reach fruition. Only once a sufficient number of modern and secular principles and structures had taken root did European society develop a strong, almost inevitable dynamic toward completing the process. Put simply, a tipping point was reached, from which there was no turning back, which ushered in the eruption era. In France, for example, we may say that the incubation period lasted from the fourteenth until the mid-eighteenth century, a time that saw the flowering of both the Renaissance and the Age of Enlightenment. The French revolution marked the beginning of the eruption phase, which lasted until approximately 1918 (or in some aspects 1945). The Jews in Europe lagged somewhat behind, with those of Western Europe embarking upon the process of modernization only in the late eighteenth century, while their coreligionists in the east had to wait, in some cases, until the mid-nineteenth century. However, for both communities, the incubation period proved to be a relatively short one, so that by the end of the nineteenth century, Jewish society throughout Europe, having reached the point of no return, moved into the age of eruption.

Thus the incubation and eruption phases vary in both duration and pace. Among the many reasons why the modernization and secularization of Jewish society took place more quickly than that of Christian Europe, one stands out above all others: Christian society was the author of its own revolution, reshaping Europe in line with its own ever-evolving modern-secular vision. Mounting such a revolution proved to be a long and arduous process, replete with trial and error, setbacks, and even defeats. The Jews, on the other hand, having become, in principle, fully functioning members of the wider community only at the end of the eighteenth and beginning of the nineteenth century, when they were finally granted their civil rights, had the advantage of entering a world that had already passed through the first stage of modernization and was now in the throes of the second. Having the benefit of both a model to emulate and a more felicitous environment in which to pursue it, they were able to pass through the initial phases of the modern revolution at an accelerated pace.

All this augurs well for the chances of launching a successful cultural revolution in the Muslim world, since, in addition to being able to look to the Christian and Jewish precedents, Muslim society, like the Jews before them, will have the bonus of landing in a modern-secular world. This will help speed up the process of secularization, which will proceed even more rapidly if it is helped along by measures rooted in past experience. Another reason for optimism is that cultural warfare is waged primarily by utilizing effectively all available systems of communication; and there is little doubt that the means of communication at our disposal today are vastly more sophisticated, numerous, and effective than those used in the past. Indeed, the proliferation of independent and informal channels of communication has resulted in some arch-conservative regimes losing their grip, in part or entirely, over the media. A prime example is the role the internet and especially Twitter played both in escalating the protests in Iran in the summer of 2009, and in bringing them to world attention.

The principal task of any cultural warfare operation aimed at modernizing and secularizing the Muslim world is to first plunge Muslim society into and then nurse it through the incubation period. Once the process has reached the tipping point and the Muslim world is thrust into the eruptive phase of the revolution, steps may be taken to nudge the proceedings along in the right direction, but ideally the whole enterprise should then be allowed to develop independently, as the product of its own internal dynamics. Accordingly, the PSYOP challenge facing the West can be more narrowly defined as one of orchestrating the incubation phase of the Muslim modernization–secularization revolution, a challenge we shall meet by using historical precedent to develop a multi-stage cultural PSYOP program.

## The Road to Modernization and Secularization: The Practicalities

The first question to ask is what prompted European society to abandon its centuries-old, and in the case of the Jews, millennia-old, rich and powerful cultures – cultures that were, moreover, imbued with the aura of sanctity. Why were these cultures abandoned in favor of newfangled, revolutionary practices, whose ultimate aim was to rid society of its tried and tested traditions? Of course none of this happened overnight, in one fell swoop; rather, it was a cumulative process that took place in several stages. We will examine each of these stages in their chronological order, describing first the Christian and Jewish precedents, and then suggesting ways in which these can be adapted and integrated into a campaign to modernize and secularize the Muslim world.

### *The First Stage: The Anti-Clerical Offensive*

During the first stage of the incubation period, the focus was on exposing what seems to be the religious establishment's endemic corruption. Targeting the clergy, making public its various indiscretions, transgressions, and even crimes, has the advantage of tacitly stigmatizing, while avoiding a direct attack on, religion itself. It is far simpler to censure and condemn individuals for their various failings, than to denounce or try to overhaul deeply entrenched belief systems. No stranger to temptation themselves, human beings find it easy to believe that even a priest, God's representative on earth, may be a less-than-shining example of moral rectitude. Naming and shaming various levels of corrupt clergy, both high and low, listing their myriad vices and villainies, exposing their hypocrisy is bound to make an impression and undermine public faith in the religious establishment as a whole.

This is precisely what Jan Hus did in the early fifteenth century when he denounced Pope John XXIII and his bishops for trafficking in indulgences.[13] Martin Luther initially directed his ire against Johann Tetzel, a priest and tradesman in papal indulgences, but he soon went on to attack the Pope, raging against Leo X's lavish building programs, which were financed largely by those same indulgences.[14] Meanwhile in England, Henry VIII, for domestic reasons, embarked upon a similar course of action, pillorying the

Catholic clergy for its rapacious nature and innate moral turpitude.[15] None of the three, however, lashed out against religion itself. On the contrary, they were themselves deeply religious, and sought to establish a more virtuous, more Christian alternative to what they perceived as a scandalously corrupt and unprincipled religious establishment. It is only later that broadsides, jeremiads, as well as satirical lampoons – such as Molière's *Tartuffe* – on the iniquities of the clergy were penned devoid of even the suggestion of a viable substitute or alternative.

As no religious establishment is entirely free of corruption and hypocrisy, priests, with their dubious claim to the moral high ground, make easy prey. In the case of radical Islam, the first line of attack should be to expose its clerics' manifold base and shabby depravities. Particular emphasis should be placed on those duplicitous priests who, while busy damning the West as the embodiment of all evil, are equally assiduous in trying to avoid deportation so that they can remain in the corrupt West, basking in its myriad pleasures. The question of financial misconduct may also prove a fruitful vein to mine, given the vast amounts of money that the Jihadist movement moves around the globe, some of which finds its way into the religious leaders' pockets in the form of commissions, interest, or simple bribes. The levying of tithes is a longstanding Islamic tradition, but these are intended to fund religious or charitable causes, not finance the clergy's lifestyles. Moreover, the rules of Islamic banking expressly forbid the charging of interest or commissions. If it were revealed, especially to the poorer members of society, that some of their hard-earned cash is finding its way into the hands of their religious leaders for their own personal use – that percentages and sweeteners are routinely offered and taken – it could go a long way toward undermining the faith of the Muslim masses in its religious elite.

However, before launching such an exposé, there is a need to reorganize and restructure the various international institutions, as well as departments within the US administration, charged with monitoring international and national finance as part of the battle against radical Islam. It is essential that these bodies, which until now have tended to operate at best independently, at worst in competition with one another, start cooperating. By working together, coordinating their operations, pooling their findings, and exchanging ideas, they will be able to function more efficiently and to greater effect.

Nor should the cooperation end there, but the intelligence community must join forces with media experts to establish a communications network, composed of both the traditional media (press, television, and radio) and newer media (websites, blogs, social networks) able to disseminate the information gathered in the most eye-catching, ear-catching manner.[16] In addition, and in order to hammer the message home, priority must be given to obtaining documentary evidence – in the form of recordings, photographs, and video – providing incontrovertible proof that religious leaders have accepted kickbacks, enriched themselves at the expense of the needy, engaged in sexual affairs, and above all, sent others to their death while ensuring their own and their family's safety. If possible this evidence should be made public, in the form of pamphlets, articles, television broadcasts, works of fiction, and satires lambasting the tartuffes of radical Islam; and, where feasible, it should be written and produced by local activists or front organizations, as any material, however well-documented, which is published in the West will be quickly denounced a tissue of lies.

### The Second Stage: Targeting the Poor and Dispossessed

All societies have their share of the downtrodden and disadvantaged, the Muslim world is no exception. Addressing and redressing the myriad injustices, the privation and abuse, suffered by various sections in Muslim society could prove to be both a morally principled and extremely beneficial course of action.

The desire for equality is at base, as Freud observed, a sublimation of humankind's feelings of envy, encapsulated by the question: "Why them and not me?"[17] In the past, the green-eyed monster was often exploited to undermine society's traditional hierarchies. In Renaissance Europe, however, this assumed a more subtle, bypassing form, with few, initially, thinking or daring to criticize the states' most powerful elites and obvious objects of envy: the king, the nobility and the church. Such overt attacks came only a few centuries later, when the principle of equality was already accepted. The struggles in the earlier stages focused mainly on the equal rights of religious minorities and "freethinkers". Tacitly, almost unconsciously, the rights of the young were also promoted, against the gerontocratic hierarchy of the medieval society. This change alone was enough at that time to strengthen the change-seeking elements within the society. It was to celebrate its crops a few centuries later, when the young were the spearheads of the great revolutionary movements.

The question of women's rights was tackled only once the second phase of the revolution, the era of eruption, was well under way. As for the poor, their all too natural feelings of envy and sense of gross injustice have long been exploited by revolutionaries and reformers everywhere. For example, many of the Eastern European Jewish youths who chose to pursue a secular education, join modern ideological movements, or emigrate (mainly to America, where the immigrants almost always embraced secular lifestyle) did so to escape the cycle of poverty which had trapped so many of their coreligionists.[18]

Playing on the Muslim youth's ambitions and desire for change may prove a difficult if not impossible task. Many younger Muslims have welcomed radical Islam as a revolutionary force, one that seeks to overturn their societies' hidebound social orders which seem determined to block any expression of youthful ambition. While it is true that the movement's spiritual leadership is dominated by middle-aged and old men, the young are encouraged to take an active role in the Islamic revolution and even shine center stage. Moreover, many of the Islamist movement's middle and lower-ranking leaders are young men, with a chosen few fast-tracked to join its more senior ranks. Even the act of martyrdom (*shahada*) is essentially a young person's game. It is also, paradoxically, a means of social advancement. Admittedly, the *shaheed* do not personally enjoy the more worldly fruits of their sacrifice; but their families certainly do, while they themselves relish the thought of being acclaimed as heroes and revered after death. In sum, and in light of all that radical Islam appears to offer Muslim youth in terms of personal, political, religious, and financial advancement, there is clearly little point in appealing to the Muslim youth's (seemingly nonexistent) sense of personal injustice and discrimination.

A much more promising course of action would be to focus on the discrimination suffered by Muslim women, simply by virtue of their belonging to a society in which women are considered, at best, second-class citizens; this is a tactic that the United States has adopted in Afghanistan, with some degree of success.[19] In an effort to refute the charge that women are discriminated against in Islamic societies, some radical and moderate Muslims launched a campaign, directed at their womenfolk, in which Islam is saluted not only as more caring and more protective of women, but also as having more respect toward them than the dissolute, sybaritic West.[20] Moreover, it is undeniable that under radical Islam, some women are able to express themselves, at least up to a point, and to pursue activities from which they had previously been barred.[21] The fact remains, however, that the vast majority of Muslim women, regarded as innately inferior to men[22], are the victims of the most appalling oppression. When addressing the question of women's rights and their role in society, it would be advisable to tread carefully, delicately balancing traditional Islamic moral standards and Western principles; a model validating the principle that all Muslim women, practicing or not, have the right to an independent, fulfilling life; that thus empowered, they can have a family, an education, and a career. At the same time, steps must also be taken to eradicate the many unconscionable crimes sanctioned by religion that are committed against women, with bogus and forced marriages and "honor killings" topping the list.

As part of addressing the question of the position of Muslim women in society, the cultural PSYOP offensive ought to capitalize and build on the work of local activists by helping opposition groups, like those in Iran campaigning for women's rights, to become more professional. Assistance should also be given to these organizations in publicizing the draconian restraints to which many Muslim women subject, including the severe restrictions placed on their movements and the harsh Islamic dress code. Finally, a spotlight must be thrown on the women sentenced to death by stoning for so-called sexual transgressions. At this stage it would be inopportune, if not counterproductive, to insist that the problem lies in Islam itself, in the fact that it regards women as lesser beings; rather, the sufferings of Muslim women should be ascribed to the perversions of the religious establishment which has subverted Islam for its own misogynistic ends.

Other than women, the group with the hardest lot in society has generally been the poor. Exploiting the masses' envy and resentment of the more privileged members of society can, especially when justified, pay high dividends: revolutionaries of many different cultures and periods are quick to raise the banner of economic injustice and social inequality. In fact radical Islam does this very thing, striving to prove that caring for the poor is fundamental to Islamic lore; one of the cornerstones of Ayatollah Khomeini's campaign, prior to the Iranian revolution, was its insistence on Islam's devotion to the well being of its poverty-stricken masses. There is little to be gained from disputing the veracity of this claim and becoming enmeshed in an unwelcome debate over basic tenets of Islam. Far better to attack the Muslim countries' self-serving and grasping elites, most notably the legion of civil servants who seem bent on enriching themselves at the expense of the poor. Revealing the degree to which the elite are oblivious to the fate of the poor will go far toward convincing the latter that the entire system is rotten to the core and in dire need of reform.

The wealthier Arab states, Saudi Arabia and the Gulf states, where the gap between the rich and the poor and even the middle classes is so vast as to be virtually unbridgeable, are tailor-made for such a crusade. Once the cultural warfare campaign against these countries' ruling elite has begun, it will be the turn of their allies in the clergy, many of whom have rushed to embrace their cronies' luxuriant lifestyles. The final line of attack will be to expose those apparently ascetic clerics in an effort to strip them of their hypocritical austere pretensions[23]. Any intelligence about these men's hidden riches or secret bank accounts should be made public which will slowly gather pace until the facts become common knowledge. Fabricating stories or forging documents exposing financial double dealings ought be eschewed because the truth is always more effective than lies, which can, moreover, be easily exposed, undermining the entire operation's credibility.[24]

## The Third Stage: Promoting Internal Religious Reform

The modernization of both Christianity and Judaism began with religious movements whose stated goal was to reform religion, not dispose of it altogether. In Western Europe, the reform movement was led by Protestant groups alongside several dissident Catholic denominations. Weakening the Catholic church's grip on the masses, their activities helped shatter Rome's monopoly on religion in Europe, leading eventually to the establishment of Protestant churches. As for the Jews, religious reform was high on the agenda of Western Jewish activists, while Eastern Europe saw the rise of a Jewish reform movement rooted in the principles of enlightenment, which carefully refrained from attacking Judaism itself.[25] At times, Jewish reformers enjoyed the tacit and even explicit support of the gentile authorities; this was a mixed blessing, given that it allowed their opponents to vilify them as heretical turncoats. Yet even these damning imprecations failed to stop their ideas from making significant inroads in the Jewish community.

At the center of all religious reform lies its challenge to the establishment's monopoly over the interpretation of the community's holy texts. As canonic texts are open to interpretation, reformers argue that their readings most accurately reflect the earliest precepts of their faith; precepts that marginalized, perverted, and ignored by the religious establishment, are ripe for revival. Since some reform movements are more radical than others, the question naturally arose, especially among the traditionalists, as to the limits of legitimate change, a question that the religious establishment was also eventually forced to wrestle with.

Despite the Islamists' attempts to present Islam as a monolith, it is replete with competing interpretations, schools of thought, and reform movements, not to mention rival sects, some of which – such as the Ahmadiyah, the Isma'iliyah and Sufism – boast a long and distinguished history. Today, young Western Muslims such as Irshad Manji who are campaigning for a thorough overhaul of Islam insist that it adapt to the modern world by embracing such values as openness and gender equality.[26] The cultural warfare program should highlight Islam's diverse nature and history and help give voice to reformers, with one proviso: support should be given discreetly so as to avoid besmirching the reformers with the taint of collaboration. Fostering a demand from within Muslim society to rekindle the forgotten rational principles that shaped Islam at its inception will, hopefully, set off a general rationalist revolution in Islamic thinking though caution should be exercised to hinder the fundamentalists insert their violent notions.[27]

### The Fourth Stage: Freedom and the "Good Life"

In the nineteenth century, the only way most young Jews could hope to improve the quality of their lives was simply to leave their communities. Compared to the life they left behind, the world beyond offered them untold opportunities on the professional, financial, and personal levels. Released from the suffocating burden of religious duty, they were at liberty to pursue the "good life", including a healthy and fulfilling relations with the opposite sex. Fearful of humankind's sexual nature, traditional, conservative, and religious cultures strove, and strive, to limit all sexual expression. As a result, many of those who embraced modernity – particularly if they were in the throes of puberty – were undoubtedly seduced by the prospect of being able to freely and guiltlessly indulge their natural urges.

Ideally, given the sensitive nature of all such delicate matters, those interested in the modernization of the Islamic world would promote a process in which it will be offered a choice of secular attitudes to relations between the sexes, ranging from the conservative to the permissive. In the past, the Christian-Arab template for such relations could have been presented as a model worth following; unfortunately, doing so today would, especially within the Arab community, be at best ill-advised and at worst disastrous. Instead, the focus should be on exposing the failures and hypocrisies of the prevailing Islamic stance on sex. There is a Janus-like dualism in Islam's attitude regarding sex: on the one hand, in public there is a close-minded, uncompromising puritanism, and on the other, and behind closed doors there is an almost prurient libertinism. A case in point is the Shi'ite institution of short-term marriages. Originally designed to resolve the problem facing warriors who, on the eve of battle, had neither the time nor money to partake in lengthy marriage ceremonies, this custom was perverted by the Hezbollah, who, seeking to raise the morale of its fighting men, gave its blessing to short term marriages (literally overnight affairs) that it has not surprisingly been accused of pimping in the name of religion.[28] Making public the religious establishment's attitude regarding sexual matters, and exposing this type of behavior, would be a first step toward introducing young Muslim men and women to a more liberated, albeit moral, environment: a world in which they can mix freely together, released from the scourge of religious supervision.

A word of caution: this stage of the PSYOP campaign should be launched only once the previous value-based stages have taken root. Promoting the good life and especially sexual freedom without the necessary moral and ideological buttresses would provide ammunition to Islam's conservative forces. No stranger to attempts to woo away its young, to bedazzle and seduce them with the delights of modern Western life, the Islamic religious establishment has developed an impressive array of counter-measures. It can and has argued that these visions of the good life are the work of the devil; that they are bereft of any moral or ethical content and seek to legitimize and give free reign to humankind's baser impulses, impulses that must be checked should one wish for redemption. Unless messages on this subject are grounded in a persuasive moral value system, many Muslims will instinctively accept praise of the good life as the embodiment the West's wanton decadence.

Making overt and covert efforts to promote a particular way of life is precisely what the Soviet Union did in Western Europe during the Cold War. Unfortunately for the Soviets, their efforts came to nothing largely because there was a much more attractive alternative on offer. Forced to choose between the enlightened, liberal, and prosperous lifestyle advanced by the United States and the rigid, ideological, and spartan one offered by the Soviet Union, most Western Europeans plumped for the former. Choosing between the two became easier still once news of the repressive nature of the Soviet regime began to trickle through. The challenge facing the West today in persuading the Muslim world to embrace its way of life is both simpler and more complex. While the material attractions of the modern Western world are indisputable and have multiplied over time, there has also been, certainly among the young and especially in the West, a growing dissatisfaction with what is seen as a culture of mindless consumerism and a hankering for a simpler, more value-based way of life. One way of squaring the circle is to offer young Muslims a progressive, liberal but ethically grounded social and economic lifestyle. The *Fatwa* (religious edict) could be issued in a desired way coupled by rulings and novel interpretations of the oral tradition that bridges between Islamic values and modernity. Islam though conceived in the seventh century has no intention after all to keep twenty-first century believers in tents and caves.

Fortunately, thanks to the modern media – radio, television, and the Internet – the Muslim world is not entirely ignorant of the merits of the good life. Muslim immigrants, and especially young Muslims students studying abroad, have also helped acquaint their compatriots back home with what the West has to offer. Unfortunately, thus far the West has failed to exploit these opportunities to launch an effective public relations offensive. This is partially because, inclined to regard its Muslim immigrants as guests, and unwanted guests as that, the West is generally quite happy to see the back of them as soon as possible. Conversely, post-colonial guilt has also played its part by inhibiting any attempt to seek a *quid pro quo* from either Muslim immigrants or students in return for Western hospitality. The PSYOP campaign, rising above these two contradictory prejudices, should ensure that formal arrangements are made guaranteeing a range of benefits to any Muslim returning home, even if only for a while, to trumpet the glories of life in the West. Concurrently, a stake should be acquired by the campaign organizers in newspapers, television, and radio stations in Islamic countries in order to gain a measure of influence over their content and emphasize the celebration of Western lifestyles. Opinion makers within Muslim society ought to be offered various incentives, bonuses, and sweeteners if they exercise their influence in the right direction. The target of this PR offensive is less those sections of society already convinced of the merits of the good life, and more the uncommitted and old conservative elite.

While Muslim immigrants enlighten their compatriots as to the sundry benefits of the good life, other more indirect, even subliminal, methods can also be used to convert the masses to the credo of Western living. Television drama or sitcoms, starring ordinary Arab-American families, could showcase homes with the latest gadgets, well-stocked fridges, bathrooms and cars for all. American universities and colleges could offer courses on Hollywood, which coincidentally celebrates the American way of life. But for this message to truly infiltrate the masses, the West must above all exploit the internet, ensuring that all people, especially in the Arab world, have access to it. This means investing in both infrastructure and local high-tech companies. It also means providing the less privileged members of society with cut-price laptops and access to subsidized computer centers, as well as teaching them basic computer and literacy skills. Admittedly, radical Islam has made excellent use of the internet to gain adherents. But the number of websites promoting the Jihadist message are a mere drop in the internet ocean compared to the vast number of secular websites awash with the promise of the good life.[29]

### *The Fifth Stage: Cultivating the Habit of Critical Thinking*

Having laid the groundwork in the previous stages, the PSYOP campaign can now move on to perhaps the most critical phase of the operation, which is the very heart of the secularization process. In Christian and Jewish Europe, the secular revolution was spearheaded by the brave few who dared to challenge not merely the traditions and institutions of organized religion, but also the very foundations of religious faith itself. Branded dangerous subversives, abused, persecuted, and even martyred, these revolutionaries were canny enough to imbue their trials and tribulations with a romantic heroic aura. This undoubtedly helped them promote their ideals, and as modernization and secularization gathered pace, they were joined by a growing number of converts to the cause.

In intellectual terms, this part of the secular revolution can be broken down into three stages. First, the historicization of religion, in which religion is redefined as a social-historical phenomenon that is, like others of its kind, rooted in and shaped by the world around it. During the second stage, religious texts, no longer sacrosanct, are subjected to a rigorous literary, philosophical, and social critique. These two stages lead to the third, when the myriad contradictions between religion, on the one hand, and science, morality, and the needs of a modern progressive society, on the other, are paraded for all to see.

Given the current situation in the Muslim world, the chances of a similar intellectual revolution sweeping the region are minute, making this possibly the most difficult and problematic stage of all. Yet, should the preceding stages of the campaign be executed in a proper and orderly manner, the odds of such a radical conceptual upheaval taking place will increase dramatically. On the other hand, launching the fifth phase of the cultural warfare campaign before the conditions for it are ripe would be ill-advised, if not calamitous. If the West acts too soon, the opposition galvanized into action, will seize the opportunity to make its case, with many Muslims intuitively endorsing its anti-secular polemics. This will make the fundamentalist movement all the more difficult to thwart, setting back the cause of modernization years if not decades.

If this stage of the operation is to succeed, the Muslim community must first possess a well-educated elite, one that is not just economically astute or technologically savvy, but also conversant in the fields of social science, humanities, and the arts. While an educated population or even elite does not guarantee a modern secular state – as the example of Iran shows – it is nevertheless a necessary condition for the development of one. Accordingly, the focus of the PSYOP program at this stage should be on cultivating such an elite, with Western cultural institutions, schools, and universities opening branches and affiliates in Muslim countries. At the same time, money should be injected into local educational systems, using the influence obtained by donations to ensure that they adopt Western-style curricula. There is no need, at this point, to offer the entire Muslim population a fully rounded higher education. Rather, the aim should be to create a small and audacious elite to act as a revolutionary vanguard. This vanguard would promote various aspects of post-modern thought, including cultural relativism and the questioning of time-honored conventions. Although its efforts might be limited at first to translating Western tracts and novels and putting on plays and exhibitions originating in the West, this newly formed intelligentsia would eventually move on to produce similar works of its own. In either case, it will need to establish small, independent platforms from which to showcase its work. Its initial appeal would be mainly to the intellectually curious, but it should eventually attract a wider audience through the popular press and internet.

The United States pursued a similar if less ambitious program in Europe following World War II. Its task however, was made much simpler thanks to its flourishing economy, which underpinned and validated its liberal value system, the two together encapsulating the American way of life. It was a lifestyle made doubly attractive owing to Europe's mostly wretched standard of living, the result, in part, of the destruction of much of its infrastructure. The United States launched an all-out cultural offensive, founding cultural centers, staging concerts and exhibitions, and financing publishing houses.[30] As for the Soviet Union, it spent vast amounts of money trying to influence the Western media and opinion-makers. Yet its aim in investing in or buying up magazines, journals, and radio stations, as in other similar activities such the cultivation of Marxist and leftist intellectuals, was primarily political; hence also its support for organizations such as the European Nuclear Disarmament(END).[31]

Economically and physically, the situation in most Islamic countries today is comparable to that of Europe after World War II. Much of their poverty-stricken population ekes out a living in substandard conditions, without access to basic amenities of modern life, and their minuscule middle classes struggle to make ends meet: these conditions bode well for the success of this stage of the cultural warfare campaign. Nonetheless, to date, most efforts, including indigenous ones, to establish secular movements in the Islamic world have largely failed. This, though unfortunate, at least makes it possible to learn from and so avoid past mistakes. The first thing to note is that it is vital that the entire process take place gradually and incrementally. Second, it would be a gross mistake to instigate the ideological revolution with undue haste. Third, any change must appear to be the result of an internal dynamic rather than imposed on society from outside. All this will require huge financial resources, to be allocated judiciously only to the most trustworthy local operatives, who will remain to all appearances independent; no effort should be spared, however, in steering their activities, behind the scenes, in the right direction.

There is, however, no need at this stage for excessive secrecy; rather, discretion should be the order of the day. Investment in the internet ought to continue apace, turning it into an omnipresent channel of communication available to all. Now would also be the time to promote websites that validate both Arab-Muslim and Western values by emphasizing their common points and shared roots. Drawing on the historicization of Islam, these sites could underline just how much the West owes Islam: how Muslim scholars in the Middle Ages labored to preserve for posterity Greek science and philosophy, the cornerstones of much of modern Western civilization. And while high culture and philosophical tracts have their place, more popular, indeed populist, material should not be ignored, but exploited to serve social or political ends.

The optimum method of spreading these ideas, and of ensuring that they reach all levels of society, is through a pyramid-like structure, at the top of which perch the intellectual elite; they, though few in number, will develop these ideas in their abstract, academic form. The rank below will be occupied by a larger group of slightly less lofty intellectuals and educators, who will simplify the message, making it accessible to the wider literate public. The third tier is peopled by PSYOP and PR specialists adept at recasting these notions into popular, easily understood forms. Finally, at the very bottom lie the media, advertising agencies, and assorted business and commercial enterprises who can serve up these ideas in pre-digested, bite-size pieces. Given that the news and entertainment programs are a highly effective means of persuasion and indoctrination, these last two groups are no less important than the first. Indeed, bright and talented young men and women with academic ambitions should be encouraged to turn their hand to journalism, drama, and, no less importantly, satire and comedy. Having the news and shows written and produced by locals steeped in the local culture and language is also a good way of avoiding any unnecessary, imprudent gaffes.

## The Sixth Stage: Propagating Dissident Movements and Secular Ideologies

Only after the previous stage has come to fruition, only after the various ideas, values, indeed, whole new way of thinking has finally taken root, a necessarily long and laborious process, can these finally be turned into social and political currency, launching the sixth and final stage of the PSYOP campaign. In Europe, nineteenth-century nationalism, Marxism, and liberalism were the product of the earlier Age of Enlightenment; their aim was to replace the existing social-political system with other more just and equitable ones. The Jews, lacking a state of their own, sought mostly to improve their lot, and especially their legal standing, in society. By the late nineteenth century, however, with some in the community swept up in the nationalist fervor that engulfed Europe, Zionism was born. But whatever their individual goals, these diverse movements had one thing in common: they all transcended the mere political. Their aim was to transform mankind and set in motion an intellectual and cultural revolution that would, *inter alia*, minimize and eventually dissolve all forms of temporal religious authority.

These variegated secular movements shrewdly presented themselves as the *dernier cri*. Being the very essence of all that is modern and cutting-edge, they were, their supporters boasted, the inevitable product of the march of time. Traditional institutions and systems, by contrast, were dismissed as outmoded and obsolete and bound to disintegrate of their own accord. Once democracy, communism, and fascism took center stage, the battle against the more orthodox political and social regimes was temporarily set aside, as the three battled each other for supremacy, a battle from which liberal democracy ultimately emerged victorious. The result: democracy together with two of its central tenets – religious freedom and the separation of church and state – became one of the defining values of the modern world.

There is little point in introducing the full range of modern political creeds into the Muslim world. For one thing, ethical arguments aside, fascism and communism have essentially fallen by the political wayside. They are no longer realistic options, and it is reasonable to assume that neither will make more than a token appearance in Muslim society, especially if the previous stages of the cultural PSYOP offensive are carried out effectively. For another, the entire aim of the campaign is to establish democracy and democratic institutions throughout the Muslim world. This means encouraging the new intellectual elite to embrace the principles of democracy not only in word but also in deed; so that just as democracy once stood at the vanguard of the intellectual revolution, it will now stand at the forefront of the political one.

With much of the groundwork already laid, what is left is to celebrate the merits of democracy through local, regional, and global platforms. In the Muslim world this means harnessing existing as well as newly launched academic and mainstream journals to the cause. Authors applauding the democratic way of life should, like their publishers, be offered encouragements. And once published, this material should receive as much publicity as possible. Articulate and rebellious students, in need of bursaries and grants, could help spread the word of democracy among their peers. Their teachers, too, could espouse the democratic ideal, with the West funding academic fellowships and chairs to secure their continued employment and promotion. In more remote regions, tribal rivalries could be exploited by suggesting that a tribe will increase its local standing by adopting democratic structures. Finally, Muslim governments that genuinely embark on the road to modernization and democracy ought to be rewarded by helping them increase their popularity at home and raise their standing abroad. This could be accomplished by reporting how these regimes have improved their citizens' lives by giving them access to better health care and education systems, for example; the underlying implication being that those living in democratic nations are better off, better fed, better dressed, and are both healthier and happier.

The overall aim of the campaign should be to create a new, affluent, and intellectually vibrant middle class. Undoubtedly at first various new cliques and networks will surface, but hopefully they will use their power to spread the democratic message. What should be avoided at all costs is the easy option of simply throwing large sums of money about, as this will only produce a small, insular kleptocracy bent on using these monies for its own ends. Once a genuine indigenous democratic movement emerges, it will in all likelihood make significant inroads in key Muslim states, triggering an internal dynamic that will catapult the entire Muslim world to the point of no return and into the second phase, the eruption period of the modernization process. And it will be at this point that the Bush–Sharansky doctrine can finally come into its own, and democratic government can be exported to the Islamic world.

## Limits and Qualifications: The Pace and Scope of Change

Before concluding, several important points regarding the cultural PSYOP campaign's goals and techniques ought to be clarified, not merely to ensure the campaign's success but also to inject a strong dose of realism into the discussion and so avoid the danger of excessive expectations. The danger is that early disappointment may result in jettisoning the whole notion of cultural warfare and the loss of an invaluable tool in the war against radical Islam.

First, as emphasized above, the entire process of modernizing and secularizing the Muslim world must take place gradually and incrementally. It is vital that each stage of the campaign be allowed to take root before the West moves on to the next. Moreover, rather than being simply imposed on the population from above, the entire process should appear to be an indigenous one, the product of a grassroots dynamic. All previous attempts to forcibly modernize and secularize the Muslim world have failed, in part or entirely: to wit, Atatürk's secular revolution in Turkey and the Shah's "White Revolution". Impatient and overly hasty, they sought to impose a policy of modernization cum secularization almost exclusively by legislation. Moreover, by targeting only the more obvious superficial accouterments of modern life without first launching a long-term and painstaking program to transform their people's deep-seated belief systems, they failed to ensure that even the modest successes they had would endure.[32]

Second, for such an ambitious campaign to succeed, both its authors as well as those managing it must be conversant in all aspects of Islamic and Arab culture; and their efforts to come to grips with it must not cease once the operation is under way. Mastery of enemy culture is the key to all PSYOP campaigns: it is the only way of discovering which buttons to push in order to wield maximum influence. Knowing the enemy proved relatively easy during World Wars I and II, at least in Europe, where both sides shared the same Christian heritage.[33] This was also the case during the Cold War, for much the same reason. Tackling Arab and Muslim culture, however, has proved more difficult and not just because of its alien nature. The cause of the problem can be traced to the tendency of Arab intellectuals to malign Middle East Studies as condescending and demeaning. The charge is that these studies, contemptuously labeled "Orientalism", are no more than a modern-day manifestation of Western colonialism which found a ready ear in a West beset by guilt for its imperialist past. As a result, any impartial, let alone critical, investigation of Arab culture and history is often dismissed as Orientalism: biased, bigoted, and even racist. Fettered by the boundaries of the politically correct, Western understanding of Arab and Muslim culture has become severely limited, if not skewed. This is true not only in America, but even in Britain, which, thanks to its once extensive empire, could previously boast some familiarity with the Arab and Muslim world.[34] Both nations, but especially the United States, must invest time and money to make good this lacuna, with special emphasis on understanding the cultural underpinnings of the Arab education system.

Third, cultural phenomena do not develop in a void, but in conjunction with the more material aspects of daily life. Hence, any cultural changes in the Muslim world must be accompanied by sustained economic and technological growth.[35] As countries that already boast even a few modern infrastructures will, in all probability, prove more susceptible to secularization, it is advisable to begin the cultural offensive in these rather than in their less advanced countries.

Fourth, the process of modernization will undoubtedly proceed at different paces in different regions. It will take time until it reaches the more remote regions of the Muslim world, areas with little or no access to the internet. Moreover, it is here in these small, largely illiterate, mountain communities that resistance to modernization, let alone secularization, will most likely be highest. Ossified anti-modernists can be found everywhere, particularly among the traditional elite and clergy. Reactionary, dogmatic, and fearful of change (not least to their own positions), they will continue to fight in their corner long after the twin forces of modernization and secularization have won the day. But if the modernization campaign begins in the more heterogeneous and open-minded towns and cities, and only later moves to the periphery, the die-hard rural fundamentalists will be forced on the defensive: they will find themselves fighting for their survival within their own communities against the background of a reformed and remodeled culture. With most of their energy devoted to fending off so-called Muslim heretics and thwarting the threat from within, radical Islamists will have little time left for conducting an active campaign against the West.

Finally, the aim is not to create a secular, let alone atheist, society divorced from all things religious. Suffice it to say that as in the West, religion in Muslim countries will become a private affair, a matter between them and G-d. In the public arena, Islam's presence will be limited to rituals and symbols largely devoid of religious significance, the goal being a state along the lines of the American rather than French *laïcité* model. Finally, no longer enjoying exclusive government sanction, Islam will become merely one spiritual alternative vying for disciples on a level playing field.

The above caveats aside, there are several reasons why the cultural PSYOP offensive has every chance of succeeding. First, thanks to the Judeo-Christian precedent, there is no need to devise the campaign from scratch. Rather than blindly groping its way about by trial and error, the West has at its disposal a ready-made blueprint for action. Second, the process will have behind it the full weight of a powerful state coalition. Third, it will have access to the *sine qua non* of any successful PSYOP campaign: a range of highly sophisticated channels of communication, far superior to anything previously available. Finally, some Muslim states already posses the technological and economic infrastructure necessary for a successful cultural and secular revolution.

Certainly a degree of caution must also be exercised when assessing the chances of success of the cultural PSYOP campaign – and yet bearing in mind the obstacles to the campaign's success may paradoxically serve to increase the odds of a favorable outcome. These obstacles will include: first, just as the West when devising its battle plan may draw on past precedents, so too can the forces of Islamic tradition. They are no less experts in the flaws and failures of the contemporary secular world, and by noting where their European predecessors went wrong, they will be better able to defend themselves against the onslaught of modernity. Second, they too, especially if they are part of the ruling elite, will be able to exploit (where it exists) the state's economic and technological infrastructures to advance their goals. Finally, Islamists can and have made effective use of a range of modern communication systems. So adept have they become at exploiting the media, and especially the internet, that in the battle over the airwaves they will assuredly score a significant number of victories. To all this must be added radical Islam's ruthless and single-minded determination to win the war, by whatever means possible. This last element stands in stark contrast to West's tendency to dither about, ruminating over the morality of even its most innocuous policies. In short, the same fundamental rule applies in cultural warfare that applies in any other type of war: never underestimate the enemy. On a more positive note, the West still has the upper hand in virtually every one of the relevant parameters of a cultural warfare campaign; an encouraging fact, since, although victory is never easy or guaranteed, failing to launch such a campaign will almost certainly ensure the West's defeat in the war against radical Islam.

## Conclusions: Resuming the Drive for Definitive Victory

The prevailing mindset in today's White House seems to be one of accommodation and even appeasement. The notion of a conclusive victory, evidently considered outdated and defunct, has been dismissed out of hand. Having apparently abandoned the visionary goal of democratizing the Arab states and populations, President Obama seems content to establish some kind of *modus vivendi* instead. A pluralist by conviction, he readily acknowledges the legitimacy, even as he questions the sincerity, of other cultures and regimes, and is willing to extend conditional recognition to groups and states antithetical to all that the West stands for. Yet however liberal-minded President Obama may be, we have not yet reached "the end of history", let alone a utopian millennium in which the lion lies down with the lamb. Radical Islam, determined on winning its war against the West, has remained impervious to the new President's overtures. Nor does its influence over Muslims, at home or abroad, appear to have waned. If the overly optimistic policy pursued by the current administration fails (as we venture to predict it will), a new game plan will be needed, forcing the reconsideration and perhaps even adoption of old goals and strategies, including the Bush–Sharansky doctrine.

The Bush–Sharansky doctrine collapsed not because of its romantic belief that democracies are inherently peaceful, but because it failed to distinguish between aim and means, cause and effect. Focusing exclusively on the doctrine's goal, its proponents began to zealously transplant democratic institutions onto what was essentially alien soil. Bound to fail, this strategy's problems were compounded by the fact that missionary zeal, even in the name of democracy, is irrelevant when it comes to winning a war. In war, what counts is the sides' relative strength, the aim being to increase one's own strength while reducing that of the enemy. It is in the context of this strategic approach that democracy's weaknesses come into their own. The West should spread the word of liberal-democracy less because of its virtues – because it is the best of possible systems, dedicated to the values of freedom and human rights – than more because of its failings, specifically its celebration of the cult of "me". Democracy gives rise to a comparatively weak-minded and self-centered disposition, and certainly does not beget the kind of resilience and self-sacrifice needed to win hard-fought and protracted battles. This is why exporting democracy and a modern liberal temperament to the Muslim East will go far toward helping win the war against radical Islam.

In ignoring the vital link between a policy's aims and means, the Bush–Sharansky doctrine, however commendable its goals, largely neglected the problem of how to realize them. There is no shortage of academic tracts that examine the conditions required for democracy to take root in society.[36] These conditions essentially boil down to the need to construct the appropriate cultural, economic, and social foundations, without which there is little chance of establishing a stable, long-lasting democracy. Yet, the architects of the Bush–Sharansky doctrine, oblivious to or dismissive of academic theory, ignored this point. This failing led them to confuse cause and effect, assuming that once blessed with democratic institutions, Muslim society would rush to adopt the liberal secular values prized by the West. In reality, in order for democracy to take root and flourish, it is vital to first set the stage by creating the appropriate liberal cultural mind-set.

If the goal is democracy, the means to achieve this is first to suffuse the hearts and minds of the Muslim population with modern secular values, for without the necessary social, economic, and cultural sustenance, democracy is unlikely to take root. This means launching a cultural revolution, and replacing the present closed, tradition-bound cast of mind with a new broad-minded and enlightened outlook. This, in turn, will lead to the establishment of a modern, free, and secularized society, much like the liberal and even libertarian ones found in the West today. Engendering such a comprehensive, cataclysmic metamorphosis is no easy feat; but neither is it impossible, if undertaken in a slow, orderly, and incremental manner, as described in the six-stage plan sketched above.

This plan, far from comprehensive or conclusive, simply points in the right direction, offering a practical framework for action plus an assortment of measures to help ensure each stage's success, all of which can and should be developed further.

Some may accuse us for Machiavellian approach. We will not deny this accusation altogether. Machiavelli's contribution to political science was not in his immoral approach but rather in his a-moral approach. He did not deny the validity of moral values but rather advised that first we examine how things work in reality and only then see how we can achieve our goals (including the moral ones) within its frame. Following Machiavelli's advice, we have penned a practical plan of action, based on the etiological principle of social cause and political effect. This plan, which reflects the true relationship between aims and means, will allow the West to achieve its goals of democratizing the Muslim world and of winning the war against radical Islam.

The current US administration's pluralist outlook and its apparent endorsement of cultural relativism have led it to embrace, or so it seems, a policy of accommodation and appeasement, making it unlikely that it will engage in cultural warfare, let alone pursue the plan of action sketched above. Add to this its understanding of PSYOP as no more than a means of advancing tactical or strategic goals on the battlefield, and it is questionable whether cultural warfare will even make it on to the US government's agenda, let alone become a subject of serious consideration. Hence the need for a public debate on these issues and the reason why we elected to go public rather campaign for the doctrine behind the scenes, despite the possibility that by doing so we are also alerting the enemy to the dangers of this potent new strategy. This is a risk we deem worth taking, as any steps adopted by radical Islamists to combat the cultural PSYOP offensive can be overcome. Less easy is the task of changing the Obama administration's strongly held, and dare we say, antediluvian preconceptions. But it can be done, particularly by launching a vigorous public debate on the subject of cultural warfare. It is a debate that will, hopefully, lead to growing grassroots support for the doctrine and persuade the Obama administration to consider and adopt cultural warfare as a means to ultimate victory.

# Notes

## 1 The Threat: Radical Islamic Organizations

1 In 2004 the Islamic militants' principal goal was to break up the Western coalition in Iraq. Hence the devastating bomb attack in Madrid, which claimed hundreds of victims and led to Spain's announcement that it was evacuating its soldiers from Iraq. Japan, Italy, and Poland, having seen several of their citizens kidnapped by insurgents in Iraq, declared that they too were considering withdrawing their forces from Iraq. In 2010 similar pressure is being directed against the Dutch forces in Afghanistan.
2 On 30 April 2003 a British national with a perfect British accent managed to enter Mike's Place, a bar in Tel Aviv, a city well-known for its vigilance, without raising the slightest suspicion. He then blew himself up, killing three and wounding dozens of others. Since then a number of Westerners who converted to Islam were arrested by US and European security organizations. A curious case is Yahya Gadahn, born Adam Pearleman, who joined Al-Qaeda and stars in some of their conscription tapes. See http://www.msnbc.msn.com/id/11497264/.
3 A. Ponsonby, *Falsehood in Wartime* (New York: Dutton, 1928), p. 18.
4 On US PSYOP campaigns during World War II, see D. Lerner, *Psychological Warfare against Germany: The Sykewar Campaign, D-Day to VE-Day* (Cambridge, MA: MIT Press, 1971). On British PSYOP, see P. M. Taylor, *British Propaganda in the Twentieth Century: Selling Democracy* (Edinburgh: Edinburgh University Press, 1999).
5 Totalitarian states have no such misgiving regards PSYOP. The Soviet Union and its allies set up full-blown PSYOP organizations in times of peace and war. See R. H. Shulz and R. Goodson, *Dezinformatsia: Active Measures in Soviet Strategy* (New York: Pergamon Press, 1984). The Arab states, too, do not hesitate to exploit information for political advantage. Most Arab states have ministries of information, which, exercising absolute control over the media, ensure that they serve the regime's needs.
6 See the US Department of State, *Active Measures: A Report on the Substance and Process of Anti-US Disinformation and Propaganda Campaigns* (Washington, DC: US Department of State, 1986).
7 In the United States the field of Information Operations is currently handled by several organizations: the State Department handles the former United States Information Agency (USIA), what is now called

# NOTES

"public diplomacy", is in charge of distributing information to countries abroad; the CIA handles all black PSYOP; and the army is responsible for circulating information to enemy soldiers and citizens.

8 FM 3-13 (formerly FM 100-6), *Information Operations: Doctrine, Tactics, Techniques, and Procedures* (Washington, DC, November 2003), available at www.fas.org/irp/doddir/army/fm3-13.pdf.

9 For a critical analysis of Said's work see Ibn Warraq, *Defending the West: A Critique of Edward Said's Orientalism* (Amherst, NY: Prometheus Books, 2007).

10 P. Knightley, *The First Casualty: The War Correspondent as Hero and Mythmaker from the Crimea to Kosovo* (Baltimore: Johns Hopkins University Press, 2002). Ironically, the famous photo was taken some time before, and was the result of an American attack on an Iraqi tanker.

11 This was one of the reasons why the US army assumed responsibility for the welfare of the civilian population caught up in the fighting. It dropped medical and food supplies directly into the battle zones, an operation which in itself constituted a PSYOP message, proving that the United States had the locals' well-being at heart. These supplies were accompanied by PSYOP leaflets which informed the Afghani population of various other welfare measures taken on its behalf. Similarly, prior to parachuting supplies into the war zone, the Americans dropped fliers asking the peasant population to remain in place and until the mission was complete so that they would not be hit by a wayward container.

12 The US armed forces, in an effort to forestall any criticism of their activities, published on the internet many documents, particularly on its welfare policies, which the public could access freely.

13 Tom Regan, "US Image Abroad Will 'Take Years' to Repair", *Christian Science Monitor*, 9 February 2004.

14 There is much to be gained from studying the enemy's PSYOP activities. Alexander George, in his study of the Psychological Warfare Division, the SHAEF unit charged with analysing German propaganda during World War II, discovered that in most cases the unit's conclusions were spot on. Thus, prior to embarking on a military operation, the Germans would, as the unit correctly predicted, first launch a PSYOP offensive in the area, thus preparing the ground, psychologically, for attack. See Alexander George, *Propaganda Analysis: A Study of Inferences Made From Nazi Propaganda in World War II* (Evanston, IL: Row, Peterson, 1959).

15 The US came to realize the importance of consolidation PSYOP in the wake of its experience in Vietnam. In Afghanistan the US made an effort to deploy such PSYOP doctrine; this alarmed the Taliban, which attempted to countervail Western success in this area. The negligence

to include consolidation PSYOP in the plans for the Second Gulf War in Iraq is continuing to damage Western forces there as well as Iraqi society.

16   Sentient World Simulation (SWS) is an ambitious attempt to simulate combat. Its implications on PSYOP are extraordinary. See for example Mark Baard, "Sentient World: War Games on the Grandest Scale", *The Register*, 23 June 2007; and Tony Cerri and Alok Chaturvedi, "Sentient World Simulation (SWS): A Continuously Running Model of the Real World: A Concept Paper for Comments" (2006), available at http://www.krannert.purdue.edu/academics/mis/workshop/AC2_100606.pdf.

17   During World War II he United States managed, in the face of widespread doubts, to convince Japanese soldiers to surrender. It did so by addressing these soldiers' keen sense of honour, persuading them, via PSYOP messages, that rather than crying quits and falling captive they were honourably surrendering. In order to increase the credibility of these messages, the Americans channelled them through Japanese POWs rather than the Japanese immigrants fighting in the US Army (*Nisei*), who were despised by most Japanese as shameful traitors.

18   One Palestinian suicide bomber was found to have encased his penis in plaster in order to ensure that it would remain intact and fully functional once he reached (sexual) heaven.

19   See B. K. Freamon, "Martyrdom, Suicide, and the Islamic Law of War: A Short Legal History", *Fordham International Law Journal* 27 (2003), pp. 299–369.

20   Though communist Russia was the first country to use black PSYOP in the early 1920s, it was the Nazis, in the period prior to World War II, and Britain, during the war itself, who turned it into a fine art.

21   Black PSYOP techniques include bogus radio stations which broadcast descriptions of widespread (but often imaginary) anti-government agitation and demonstrations. The Americans and Soviets operated such stations during the Cold War. The United States used them both in Vietnam and during the First Gulf War.

22   An example of the amateur radio enthusiasts' efforts is www.clandestineradio.com, which records the frequencies used by suspect radio stations, and assesses whether these stations are counterfeit or genuine.

23   At present, fundamentalist organizations use either overt (white) or semi-covert (grey) propaganda to recruit Muslims living in the West to the cause. Most of this activity is conducted on the internet and in Arabic, making it very difficult for the West to monitor; not least because of the dearth of Arabic speakers among Western intelligence analysts.

## 2 The Palestinian PSYOP Campaign against the Neutrals during the Al-Aqsa Intifada: A Success Story

1 The Palestinians accomplished a significant PSYOP victory at the very beginning of their campaign by persuading both the international and Israeli press to adopt the name "Al-Aqsa Intifada", making use of the fact that what sparked the initial rioting was the visit of Ariel Sharon (then the leader of the opposition) to the Temple Mount, where the Al-Aqsa Mosque is situated. Thus Israel could be blamed for the entire affair. As the Al-Aqsa mosque is one of the holiest sites in Islam, the name also added a new religious dimension to what had hitherto been principally a political conflict. It helped galvanize the Palestinian public into action, and also garnered support in the Arab and Muslim world at large against Israel's jurisdiction over Palestine.
2 An address given by General Moshe Ya'alon at a conference in Herzliya, quoted in M. Eisen, *Imut Mugbal* [Low-Intensity Conflict] (Tel Aviv: Maarachot, 2004), p. 350.
3 For studies of the conventional aspects of the conflict see S. Eldar, *Aza Kamavet* [Eyeless in Gaza] (Tel Aviv: Miskal, 2005); A. Lev, *Betoch Ha'kis shel Ha'rais* [Inside Arafat's Pocket] (Tel Aviv: Kinneret, 2005); R. Bergman, *Veharashut Netunah* [Authority Granted] (Tel Aviv: Miskal, 2002); A. Harel and A. Issacharoff, *Hamilchama Hashviit* [The Seventh War] (Tel Aviv: Miskal, 2004); and Y. Meital, *Shalom Shavoor* [Broken Peace] (Jerusalem: Carmel, 2004). A striking exception is Stephanie Gutman's *The Other War: Israelis, Palestinians and the Struggle for Media Supremacy* (San Francisco: Encounter Books, 2005).
4 For a good general account of PSYOP see P. M. Taylor, *Munitions of the Mind: A History of Propaganda from the Ancient World to the Present Day* (Manchester and New York: Manchester University Press, 2003).
5 Yu.Ye. Serookiy, "Psychological-Information Warfare: Lessons of Afghanistan", *Military Thought* 13.1 (Jan. 2004), pp. 196–200.
6 An unofficial Israel Defense Forces definition. See Eisen, *Imut Mugbal*, 351.
7 Past military triumphs tend to reinforce this trend; conversely, past PSYOP successes, whether its own or that of others, may encourage a state to devote more resources to PSYOP in the future. The US Army has been constantly upgrading its PSYOP capabilities since the 1980s and so has NATO.
8 On this flexibility see Maj. Gregory J. Reck, "The Necessity for Psychological Operations Support to Special Operation Forces During Unconventional Warfare", *Small Wars Journal* 4 (2006), available at http://www.smallwarsjournal.com/documents/swjmag/v4/reck.htm. For an application of that principle in Lebanon, see F.M. Wehrey, "A Clash of Wills: Hizballah's Psychological Campaign

against Israel in South Lebanon", *Small Wars and Insurgencies* 13.3 (2002), pp. 53–74.
9. See Serookiy, "Psychological-Information Warfare".
10. http://www.ipcri.org/files/beilin-abumazen.html.
11. The so-called Goldstone Report is more fully the report *Human Rights in Palestine and Other Occupied Arab Territories: Report of the United Nations Fact Finding Mission on the Gaza Conflict* (United Nations Office of the High Commissioner for Human Rights, 15 September 2009).
12. When addressed to the enemy audience, the objective might be to drive a wedge deep between parties, factions, or groups within that enemy society.
13. Demonization themes, when directed at home audiences, seek to exacerbate fear and loathing of the enemy. In the Palestinian case, they succeeded to the point where a Palestinian could kill or look upon the random killing of Israeli citizens with total lack of compunction.
14. See Margaret Dudkevitch, "PA Website Posts Blatant anti-Israel, US Cartoons", *The Jerusalem Post* 5 June, 2003, available at http://www.palwatch.org/main.aspx?fi=91&doc_id=1787.
The personal attacks stopped when Sharon promoted within Israeli society the policy of retreat from the Gaza Strip during the summer of 2005.
15. See Peter Beaumont, "Shooting Back: Israeli Occupation Filmed by 100 Palestinian Cameras", *The Guardian*, 30 July 2008, available at http://www.guardian.co.uk/world/video/2008/jul/30/beaumont.palestine.
16. One of the most commonly used PSYOP tactics when denigrating the enemy is to accentuate one aspect of either his alleged or true felonious actions; another is to simply distort the facts, as will be seen later on. Such was the case when the Palestinians, exploiting the fact that the Israel Defense Forces tanks used ammunition with depleted uranium tips, accused Israel of using nuclear weapons against them. The objective facts of the case were irrelevant: what was important was the dramatic effect produced by PSYOP sound bite – in this instance, "nuclear weapons".
17. In December 2002, the Israeli ombudsman ruled that the means used by the Israel Defense Forces to disperse Palestinian demonstrations were neither lethal nor, it seems, effective. Harel and Issacharoff, *Hamilchama Hashviit*, p. 334.
18. A detailed survey titled *Experts or Ideologues: Systematic Analysis of Human Rights Watch*, produced September 08, 2009, is available on NGO Monitor at http://www.ngo-monitor.org/article/experts_or_ideologues_systematic_analysis_of_human_rights_watch.
19. Harel and Issacharoff, *Hamilchama Hashviit*, pp. 339–41.

NOTES

20 A photograph and video of one incident, in which an Israeli solder asked a Palestinian to open his violin case, received widespread publicity. The Palestinian played a tune (on his own initiative) in order to prove he was not concealing any explosives. Many were quick to draw parallels between that image and photographs of Jewish musicians playing for their lives in the Nazi death camps (an analogy the Palestinians played up in case anyone missed the point). It was a reference that dovetailed with the well-worn Palestinian theme of comparing their treatment by the Israeli government to the Jews' fate under the Nazi regime. See the BBC News report "Israel Army Forces Violin Recital", 25 November 2004, available at http://news.bbc.co.uk/1/hi/world/middle_east/4043299.stm.

21 Harel and Issacharoff, *Hamilchama Hashviit*, pp. 275–79, 370–71.

22 The concrete sections of the fence, which were designed to prevent Palestinian snipers from targeting Israeli towns bordering the Green Line, constituted a mere 4% of the entire structure, the remaining 96% being built almost entirely out of wire – a fact the Palestinian campaign failed to note. See the article on the Israeli Ministry of Defence website at http://www.securityfence.mod.gov.il/Pages/Heb/mivne.htm.

23 In a well-publicized example, the Israeli General of the Southern Command, Doron Almog, was nearly arrested at Heathrow Airport in London after an injunction was issued by human rights activists. See "Sharon Fears Arrest If He Visits London", *The Times*, 17 September 2005, available online at http://www.timesonline.co.uk/article/0,,2-1784018,00.html; and Chris McGreal, "Israeli Ex-Military Chief Cancels Trip to UK over Threat of War Crimes Arrest", *Guardian*, 16 September 2005.

24 See, for example the article in *Al-Hayat* (London), 24 July 2006, excerpted in Special Dispatch-Lebanon, no. 1219, 28 July 2006: "Columnist (and Former Editor) of *Al-Hayat*: Israel's Leaders are the Grandsons of Nazi Killers Who Assumed Jewish Identities and Fled to Israel", available at http://www.memri.org/report/en/0/0/0/0/0/0/1759.htm.

25 Many examples are documented at www.pmw.org.il.

26 From Palestinian Media Center (PMC), 13 April 2002: http://www.ramallahonline.com/modules.php?name=News&file=article&sid=922.

27 See Ali Abunimah, "Invisible Killings: Israel's Daily Toll of Palestinian Children", *The Electronic Intifada*, 10 December 2002, available at http://electronicintifada.net/v2/article957.shtml.

28 This accusation was revived later in a sloppy Swedish article accusing the Israel Defense Forces of trading in organs taken from slain Palestinians: see Saira Soufan, "Israeli Occupation Authorities Illegally Harvesting Organs of Palestinian Children", available at

http://www.ety.com/HRP/jewishstudies/spearparts.htm.
29 In November 1999, Suha Arafat, taking advantage of the publicity generated by the visit of the then US president's wife Hilary Clinton to Gaza, accused Israel in a joint press conference of routinely using poison gas against Palestinians, a criminal action that, she claimed, resulted in a great many Palestinian women and children falling victim to cancer: see Bergman, *Veharashut Netunah*, p. 49. For other similar accusations, see Bergman, pp. 46–51; and "Hillary Clinton Criticises Mrs Arafat", BBC News, 12 November 1999, available at http://news.bbc.co.uk/1/hi/world/middle_east/517983.stm.
30 This was in part due to the explosive devices employed by the Palestinians against the IDF, and in part a result of the IDF's tactic of bulldozing houses in order to clear the way for its fighting forces and reduce the number of Israeli casualties.
31 See Jeremy Cooke, "Eyewitness: Inside Ruined Jenin", BBC News, 16 April 2002, available at http://news.bbc.co.uk/2/hi/middle_east/1933300.stm; and D. Rhode, "The Dead and the Angry Amid Jenin's Rubble", *New York Times*, 16 April 2002, available at http://www.nytimes.com/2002/04/16/international/middleeast/16JENI.html.
32 D. Kupelian, "Probe: Famous 'Martyrdom of Palestinian Boy 'Staged'", *World News Daily*, 26 April 2003, available at http://www.wnd.com/?pageId=18358.
33 See the briefing by IDF Col. M. Eisin, available at http://www.israelinsider.com/channels/diplomacy/articles/dip_0204.htm.
34 Luckily for the Palestinians, blue, the official colour of all UN uniforms, is remarkably telegenic.
35 On 2 May 2002, Kofi Annan announced that the UN would disband the committee, both as result of Israel's refusal to cooperate with it (which, incidentally, gave the impression that it had something to hide, did not help the Israeli case, and played directly into Palestinian hands), and the heavy pressure exerted by the Israeli lobby on both the UN's Secretary General and the US Administration.
36 See the report of the UN Secretary-General prepared pursuant to General Assembly resolution ES-10/10, available at http://www.un.org/peace/jenin/. The number of Israeli fatalities in Jenin was 23: see Harel and Issacharoff, *Hamilchama Hashviit*, 253–68.
37 D. Raab, "The Beleaguered Christians of the Palestinian Controlled Areas", Jerusalem Letter / Viewpoints 490 (1–15 January 2003), quoting an interview with a freed Armenian monk in the *Jerusalem Post* of 24 April 2003; Raab's article is available at http://www.jcpa.org/jl/vp490.htm. See also Bergman, *Veharashut Netunah*, pp. 56–7; Harel and Issacharoff, *Hamilchama Hashviit*, p. 247.
38 The full story of this rapprochement is detailed in J. Lederman's *Battle Lines: The American Media and the Intifada* (New York: Holt, 1991).

## NOTES

39  See http://www.jmcc.org.
40  The campaign was assisted by the fact that often no mention was made of a particular report's or picture's source; it was credited to the news agency, rather than to its actual Palestinian source. See Dan Diker, "The Influence of Palestinian Organizations on Foreign News Reporting", *Jerusalem Issue Brief* 2, no. 23, 27 March 2003, available at http://www.jcpa.org/brief/brief2-23.htm.
41  Harel and Issacharof, *Hamilchama Hashviit*, pp. 165–7.
42  Israel, by contrast, boasts a mere seventy-five embassies: see http://www.mfa.gov.il/MFA/Sherut/IsraeliAbroad/Continents.
43  K. Kirisci, *The PLO and World Politics: A Study of the Mobilization of Support for the Palestinian Cause* (London: Frances Pinter, 1986).
44  For example: http://www.electronicintifada.net.
45  See http://www.birzeit.edu.
46  One such site is the internet page devoted to Jeanmarie Condon, a senior ABC producer: Details at http://www.cies.org/stories/s_jcond.htm.
47  Beside extensive media coverage, Rachel Corrie's tragic death produced a Broadway play, an art exhibition, websites, and numerous events, including a campaign against Caterpillar Inc.: see http://rachelcorrie.org/news.htm. For the obituary see Carl Arrindell, "Tom Hurndall: An Aspiring Photojournalist and Committed Peace Activist", *The Guardian*, 22 January 2004, available at http://www.guardian.co.uk/news/2004/jan/22/guardianobituaries.israel.
48  See "Rafah Marking Corrie Anniversary: Palestinian Children Have Marked the Second Anniversary of the Death of US Activist Rachel Corrie, Killed During an Israeli Army Operation in Gaza", BBC News, 17 March 2005, available at http://news.bbc.co.uk/2/hi/middle_east/4359325.stm.
49  See James Fallows, "Who Shot Mohammed al-Dura?" *The Atlantic* (June, 2003), p. 49, available at http://www.theatlantic.com/doc/200306/fallows.
50  See http://www.seconddraft.org.
51  See the Richard Landes film *Pallywood, 'According to Palestinian Sources . . . '*, available at http://www.seconddraft.org/index.php?option=com_content&view=article&id=522:pallywood-qaccording-to-palestinian-sourcesq&catid=58:according-to-palestinians-sources&Itemid=159; see also "Resurrection in Palestine", http://www.youtube.com/watch?v=xRz5WnHemkw.
52  Ashraf Khalil, "Mideast Debate Takes Root at UC Irvine: Jewish and Muslim Leaders Say That Clashes on the O.C. Campus Have Intensified", *Los Angeles Times*, 27 May 2006, available at http://articles.latimes.com/2006/may/27/local/me-muslim27.

53 Whether this was a pre-planned policy or mere exploiting of an opportunity remains unknown.
54 Eldar, *Aza Kamavet*, pp. 222–23; also James Bennet, "Israel Clamps Down on West Bank and the Gaza Strip", *New York Times*, 27 January 2003, available at http://query.nytimes.com/gst/fullpage.html?res= 9C03EEDD1239F934A15752C0A9659C8B63&pagewanted=all.
55 In 2006 the Palestinians finally voted the PLO out of office, having tired of its corrupt and inept rule.

## 3 From Oslo to Jerusalem: Fifteen Years of Palestinian Psychological Warfare against Israel (1993–2008)

1 See Jim Lederman's book on this period: J. Lederman, *Battle Lines: The American Media and the Intifada* (New York: Henry Holt & Co., 1990).
2 R. Schleifer, *Lochamah Psichologit* (Psychological Warfare) (Tel Aviv: Ma'arachot/Misrad Habitachon, 2007), pp. 173–180. See E. Navon's article where he mentions explicitly the Palestinians' copying of Chinese and Vietnamese propaganda models: E. Navon, "Soft Powerlessness: Arab Propaganda and the Erosion of Israel's International Standing", Working Paper, Herzliya Conference, 21–24 January 2006.
3 Admittedly a political advertisement in a newspaper can be part of a legitimate process of political campaign, but if it is initiated by a hostile entity and is part of a general campaign against an enemy, then it belongs to the realm of covert PSYOP.
4 See D. Garnett, *The Secret History of PWE: The Political Warfare Executive 1939–1945* (London: St Ermin's Press, 2002).
5 See http://www.mideastweb.org/plo1974.htm.
6 "The Unit that Is Going to Drive Our Enemies Crazy", (Hebrew) NRG website, http://www.nrg.co.il/online/1/ART1/020/638.html. "IDF Psychological Warfare Units at Work", A. Harel, *Haaretz*, http://www.fromoccupiedPalestine.org/node/1490.
7 On Hamas taking over the IDF radio station see "Live From Gaza, Hamas Broadcasting", available at http://www.ynet.co.il/articles/ 0,7340,L-3647731,00.html.
8 Prof. Yoav Gelber found thirty-six separate websites dealing with the massacre in Tantura in the War of Independence, even though no such massacre ever took place. "The Generals Are Returning Fire in the Internet", *Yediot Achronot*, 14 December 2001.
9 According to Chani Ziv, a researcher, Dr. Chanan Ashrawi is behind this activity. *Yediot Achronot*, 14 December 2001.
10 "The Anti-Israel Incitement on the Internet", *Yediot Achronot*, 17 April 1995, p. 6.
11 Ali Waked, "Arab Media: Shalit Wounded in Bombing", *Ynet News*, 29 December 2008, available at http://www.ynetnews.com/ articles/0,7340,L-3646198,00.html.

NOTES

12 See K. Sengupta and B. Lynfield, "Tunnels: The Secret Weapon for Hamas", *The Independent*, 6 January 2009, available at http://www.independent.co.uk/news/world/middle-east/tunnels-ndash-the-secret-weapon-for-hamas-1228140.html.

13 Palestinian organizations published a series of publications in Arabic, entitled *The Case of* . . . . The series included titles such as *The Vietnamese Case, The Cuban Case, The Algerian Case*, and they are available at the library of the Truman Institute in the Hebrew University in Jerusalem. Another popular work was the Brazilian Carlos Mariguella's *Minimanual for the Guerrilla*, published an appendix in Robert Moss's *Urban Guerrilla Warfare* (London: International Institute for Strategic Studies, 1971).

Y. Harkabi noted in the mid-970s that the Palestinians wishing to learn the foundations of Zionist success translated Chaim Weitzman's *Trial and Error* and Menachem Begin's *The Revolt* into Arabic: see Y. Harkabi, *The Arab Attitudes to Israel* (New York: Hart, 1975), p. 214.

14 See Y. Yoaz, "State Commission to Examine Civilian Deaths in 2002 Shahade Assassination", *Haaretz*, 22 July 2002, available at http://www.haaretz.com/print-edition/news/state-commission-to-examine-civilian-deaths-in-2002-shahade-assassination-1.229532.

15 Bearing in mind it was only three years after the roundly condemned attack on the Olympic village in Munich.

16 R. Gledhill, "Synod in Disinvestment Snub to Israel", *The Times*, 7 February 2006, available at http://www.thetimes.co.uk/tto/news/uk/article1948453.ece. The main mover behind the divestment campaign is the Palestinian Anglican clergyman Dr. Naim Atik: see http://www.sabeel.org/.

17 "Candidates Are Buried to See Which Will Break", *Yediot Achronot*, 17/4/1995, p. 4.

18 See "Darkness at Noon", *Solomonia*, 24 January 2008, available at http://www.solomonia.com/blog/archive/2008/01/darkness-at-noon-msm-plays-along-with-ha/index.shtml, which includes photos from the *Jerusalem Post* showing that some of the blackout photos were staged.

19 See "Gazan Doctor Says Death Toll Inflated", *Ynet News*, 22 January 2009, available at http://www.ynetnews.com/articles/0,7340,L-3660423,00.html.

20 The claim was often that in essence there are no differences between the two warring sides: the Israelis have sophisticated weapons, and the Palestinians have the suicide bombers. See http://www.mfa.gov.il/MFA/MFAArchive/2000_2009/2001/5/Palestinian%20Incitement%20of%20Suicide%20Bombings.

21 There is, of course, the risk that such images will make it easier for the enemy to go to war, but since Israel at that time was on an appease-

ment track, the danger was not very great. See the photograph of a Palestinian demonstrator wielding a large knife, his face covered with a terrifying mask, in *Yediot Achronot*, 19 April 1995. The strange, fearful mask served the dual purpose of concealing the demonstrator's identity from the IDF and the General Security Service, and of terrifying the enemy.

22. A video clip of an execution of Fatah activists by Hamas can be seen at http://video.google.com/videoplay?docid=3383195979997564234#docid=-6041204945834291260.
23. The Palestinian PSYOP targeting of American Jewry is a complex issue. To make a long story short, the Palestinian strategy holds that pressure must be exerted on the Jewish community, so that it will in turn influence Israel, with the goal of releasing the pressure on the community. Israel would then be seen as a burden to American Jewry.
24. "A Terrorist Leader Announces: We Have Trained 70 Human Bombs, Thousands of Volunteers Are Ready to Participate in Suicide Attacks", *Yediot Achronot*, 16 April 1995, p. 7.
25. The commander of the paratroopers' brigade said after the military action in Shechem that the "Balata (a refugee camp in Nablus) tiger turned out to be a pussycat." See *The Jerusalem Post*, 3 March 2002, p. 2.
26. See "Head of Research Section of the IDF Information Branch in the Knesset Committee", *Hazofe*, 26 December 2008, available at http://hazofe.co.il/web/katava6.asp?Modul=24&id=53153&Word=&gilayon=2960&mador. As mentioned above, in this incident Hamas overdid it, and the Israeli response was very severe, and in Hamas's eyes, disproportional.
27. See "Closing an Era of Enmity", *New York Times*, 24 April 1996, available at http://www.nytimes.com/1996/04/24/opinion/closing-an-era-of-enmity.html.
28. It was some years before the politicians could no longer ignore the intelligence reports of Arafat's involvement in terrorizing Israel. See M. Kalman, "Terrorist says Orders Come from Arafat", *USA Today*, 14 March 2002, available at http://www.usatoday.com/news/world/2002/03/14/usat-brigades.htm
29. "Arafat: IDF Officers Are Sabotaging the Peace Agreement", *Haaretz*, 10 January 1995, p. 2a.
30. See "Protest in London against Israel's Invitation to Turin Book Fair", available at http://www.inminds.com/article.php?id=10262; and "Thousands Protest Turin Book Fair's Israel Theme", *The Jordan Times*, 11 May 2008, available at http://www.jordantimes.com/?news=7783.
31. See "Hillary Clinton Criticises Mrs Arafat", BBC News, 12 November 1999, available at http://news.bbc.co.uk/2/hi/517983.stm.

NOTES

32 See "Hamas Leader Asks for Gaza Strip Help", *USA Today*, 20 January 2008, available at http://www.usatoday.com/news/world/2008-01-20-gaza-electric_N.htm.
33 The first to use neonatal units as a means of demonization were the Americans in the First Gulf War (1991) against Iraq: see http://www.psywarrior.com/HerbDStorm.html. The Palestinians were quick to make use of that successful idea against Israel.
34 One example of the Palestinians letting the pressure become too great was Hamas's provocation of Israel at the end of the Tahadiyeh (December 2008), firing dozens of Kassam rockets and mortar shells into its territory, as it brought about Operation Cast Lead, which began at the end of that month and lasted into January 2009. Similarly the suicide bombings during the Al-Aqsa Intifada, perpetrated mainly by Hamas and the Islamic Jihad, brought about Operation Defensive Shield (2002), in which the IDF took over the Palestinians cities in Judea and Samaria. The PA generally avoided these excesses for strategic reasons, but supported them clandestinely.
35 *Yediot Achronot*, 18 June 1995, p. 3.
36 "Hamas Denies Responsibility in the Attack", *Yediot Achronot*, 27 June 1995, p. 12.
37 "Molotov Cocktails against Tanks", *Yediot Achronot*, 13 July 2001. Photographs of the tunnels in Gaza, in which even animals are sent through, were published in August 2008 by Hamas. See "Hamas Operates 200 Tunnels and a Fuel Pipeline", at http://www.haaretz.co.il/hasite/spages/1016746.html.
38 Syria constantly emphasizes that the renewal of any negotiation with it must begin from the last point reached in past negotiations.
39 "Toned-Down Welcome to the 101 Released Prisoners", *Haaretz*, 9 January 1994, p. 2a.
40 "Rabin Turned Down a Request to Release Sheikh Yassin", *Yediot Achronot*, 27 October 1993.
41 *Haaretz*, 7 February 1995.
42 Anthropologists have written much on the culture of shame in the East vs. the culture of guilt in the West. One of the better known scholars in this field is Gert Hofstede, who conducted a global research project for IBM. While it is easy to make light of this claimed cultural difference by labeling it as racism, it is important that decision-makers examine events with a perspective beyond that of mere political correctness. See G. Hofstede, *Culture's Consequences: Comparing Values, Behaviors, Institutions and Organizations Across Nations* (Thousand Oaks, California: Sage Publications, 2003).
43 See "Heavy Civilian Casualties after Drone Attacks", available at http://uniorb.com/RCHECK/drone.htm.
44 Author's interview with an infantry officer, May 2008. When the inter-

viewee was serving a tour of duty in 2004, he was the commander of an IDF squad that went into a Palestinian house in the Gaza Strip. At midnight they spotted an armed insurgent with an RPG rocket launcher, surrounded by a group of sleepy children as protection. The squad took no action.

45  In October 2008, Yonatan Dachoach-Halevi discovered that the influential human rights organization Betzelem counts terror activists as civilian casualties in its data. See http://www.haaretz.com/hasite/spages/1031429.html?more=1.
46  *Yediot Achronot*, 29 March 1995.
47  "1500 ID Cards Captured in Gaza", *Yediot Achronot*, 15 May 1995, p. 12.
48  "Arafat: Ghandi Perpetrated the Attack", *Yediot Achronot*, 3 March 1996, p. 7.
49  E. Karsh, *Arafat's War: The Man and his Battle for Israeli Conquest* (New York: Grove Press, 2003), p. 138.
50  Over a dozen pro-Arafat biographies were written during his lifetime, mostly with his encouragement and full cooperation. His false Palestinian background was revealed only in Efraim Karsh's book (see previous note).
51  *Yediot Achronot*, 13 April 1995, p. 3.
52. "The Palestinian Authority has Arrested Hamas Leaders", *Yediot Achronot*, 27 June 1995, p. 12.
53  For instance, the Palestinians made use of Yitzchak Rabin's statement (1992) that he wished Gaza would drown in the ocean. The rest of the sentence – that since that this is not the case, we must negotiate with the Palestinians – was omitted.
54  See http://www.honestreporting.com/articles/45884734/critiques/Gaza_Beach_Libel.asp.
55  The photograph of the girl who lost an eye to a rubber bullet, which was taken by Stan Grossfeld, was published in the June, 1997 issue of *Life Magazine*, and may be seen at http://www.whale.to/b/israeli_apartheid.html.
56  *Haaretz*, 5 March 1996, p. 24.
57  The methodology of the surveys and even the wording of the questions used are generally not made public, and one wonders how the surveys are conducted in a totalitarian society such as the PA or Hamas-ruled Gaza. If one can draw conclusions from an analysis of the population data from the Palestinian National Bureau of Statistics, which found a discrepancy of one million people between reality and the bureau's statistics, it seems there is room to doubt the validity of the Palestinians' surveys. See Benet Zimmerman and Yoram Ettinger, "The One Million Discrepancy", available at http://www.biu.ac.il/SOC/besa/MSPS65Heb.pdf.

NOTES

58 For instance, a survey by the "Independent Palestinian Press Agency" which reported that "75% of the Palestinians support the peace process" was awarded (June 1995) a prominent spot and a large heading in *Yediot Achronot*. See *Yediot Achronot*, 23 June 1995, p. 6.
59 Statistical data are almost always accepted by the media unquestioned. Danny Rubinstein, "The Fatah Leads in All Surveys", *Haaretz*, 23 January 2006. http://www.haaretz.co.il/hasite/objects/pages/PrintArticle.jhtml?itemNo=673611.
60 See, for example, the legal acrobatics used by the US in order to enable Arafat to speak at the UN two years after the murders in the Munich Olympics. Arafat landed on the roof of the UN building in a helicopter, so as not to set foot on American soil.
61 On Bil'in see http://www.bilin-village.org/; on the boycott see http://www.pacbi.org/; on the Swedish blood libel see "Swedish Paper's Organ Harvesting Article Draws Israeli Outrage", CNN, 19 August 2009, available at http://articles.cnn.com/2009-08-19/world/israel.sweden.organ.harvesting_1_israeli-forces-israeli-soldiers-palestinian-man?_s=PM:WORLD.
62 See *Soviet Active Measure in the 'Post-Cold War' Era 1988–1991: A Report Prepared at the Request of the United States House of Representatives Committee on Appropriations by the United States Information Agency* (Washington, DC: United States Information Agency, 1992), available at http://intellit.muskingum.edu/Russia_folder/pcw_era/index.htm#Contents. The story spread throughout South America was that rich Americans come to purchase organs for transplants, taken from local poor children by gangs of kidnappers.
63 See "Italy-Based Group Accuses Israel of Poisoning Gaza Land", 5 January 2010, Al-Manar (Hezbollah TV), available at http://www.almanar.com.lb/newssite/NewsDetails.aspx?id=118344&language=en

## 4 Hasbara, Propaganda, and Israeli Public Diplomacy: A Historical Perspective

1 Hasbara is the Israeli term for what elsewhere has been called "propaganda" (a terms used in Nazi Germany), "Strategic Psychological Operations" or PSYOP (the old term used by the US Army), "Public Diplomacy" (a term used by the US government) and Military Information Support Operations or MISO (the latest doctrinal development of the US army). It originates from the word *Le'hasbir*, which means "to explain", and will be used throughout this chapter.
2 This issue was crucial in the First Lebanon War of 1982, for example. Today this broadcasting infrastructure is unimportant, since information is sent by email, satellite phone, etc., without any need for central satellite services.

# NOTES

3  Moshe Yegar, former assistant Director General of the Ministry of Foreign Affairs, which is responsible for the country's image abroad, described the organizational aspects of the hasbara failures in that ministry. Many articles have been written in Jewish journals, especially in the US, about the lack of communication between the establishment and the community, and the lack of understanding by Israelis of the mechanisms that operate in American public opinion. See M. Yagar, *The History of the Foreign Hasbara of Israel* (Hertzlia: Lahav, 1986) [Hebrew].
4  M. Stern, ed., *Greek and Latin Authors on Jews and Judaism* (Jerusalem: Israel Academy of Sciences and Humanities, 1974), p. 97. Mnaseas of Patra was the first to recount the story that the Jews worshipped the head of an ass, and Apion followed in his footsteps and wrote his anti-Semitic book.
5  A. Kasher, ed., *Neged Apion* [Contra Apionem] (Jerusalem: Zalman Shazar Center, 1997) [Hebrew].
6  In a debate with the minim (the early Christians), "they asked R. Simlai ... what would you say to us?" Midrash Vayikra Rabba 4:6. Here the expression is "broken reed". See Midrash Tanchuma, Chukat, ch. 8, etc.
7  On the Jewish sages' attitude to non-Jews and to Greek wisdom see S. Lieberman, *Greeks and Hellenism in the Land of Israel* (Jerusalem: Mossad Bialik, 1963).
8  A comprehensive picture of these debates can be found in the popular collection of Eisenstein titled *Otzar Havikuchim*. See Y.D. Eisenstein, *Otzar Havikuchim* (New York, 1922).
9  According to advice given by a Jewish scholar in the twelfth century, a Jew should not engage in a public dispute with Christians because in effect he will lose no matter what the outcome. See F.E. Talmage, "Christianity and the Jewish People", in F.E. Talmage, ed., *Readings in the Jewish Christian Encounter* (New York: Ktav Publishing House, 1975), pp. 240–253.
10  Eisenstein, *Otzar Havikuchim*, pp. 20–21.
11  Schur argues that the Karaites did not take over mainstream Judaism due to the work of Rabbi Saadia Gaon (882–942). See N. Schur, *History of the Karaites* (Frankfurt: Peter Lang Publishers, 1992), pp. 53–54. Karaites and Christians cooperated against mainstream Judaism. See D.L. Lasker, *From Judah Hadassi to Elija Bashyatchi: Studies in Late Medieval Karaite Philosophy* (Leiden: Brill, 2008), pp. 190–191.
12  Such as the works of Rabbi Saadia Gaon in *Emunot Ve' Deot* (Beliefs and Knowledge) and the translation of the Torah into Arabic to make it accessible to common people, together with books on grammar and a prayer book.

13 Rambam (Maimonides), *Hilchot Mamrim*, Ch. 3:1–3 (Hebrew).
14 A. Altmann, *Moses Mendelsohn: A Biographical Study* (Philadelphia: Jewish Publication Society, 1973).
15 Moses Mendelssohn, *Moses Mendelssohn: The First English Biography and Translations*, ed. James Schmidt (Bristol: Thoemmes, 2002); Mendelssohn's Jerusalem is in vol. 3 of this collection.
16 Y. Katz, *Out of the Ghetto: The Social Background of Jewish Emancipation 1770–1870* (Cambridge, MA: Harvard University Press, 1973), pp. 51–53.
17 See Katz, *Out of the Ghetto*, chapter 11.
18 E. Benamozegh, *Israel et l'humanité* (Paris: Michel, 1961).
19 On the Noachides see http://webpages.charter.net/chavurathbneinoach/moshiach.html; on Noachide teaching and practices see http://www.asknoah.org/Academy.html.
20 Rabbi S.R. Hirsch, *The Collected Writings* (New York: Feldhaim, 1984).
21 According to Maharam Shik on *Pirkei Avot* 2:14 ("Know what to answer the heretic"), one must answer such a Jew only when there is a chance he will repent. If there is no chance, or if it is a distant chance, one should not spend time on it at the expense of Torah learning.
22 Y. Katz, *Hatred of the Jews, from Hatred of the Religion to the Negation of the Race* (Tel Aviv: Am Oved, 1979), pp. 7–14 [Hebrew].
23 Rashi, *Parashat Vayishlach* on Genesis 33:4, following Bereishit Rabba.
24 See one strident expression of this approach in the twentieth century by Rabbi Yekutiel Yehuda Halberstamm of Sanz-Kleusenburg in *The Way We Go, Chapters of Education and Guidance* (Igud Chasidei Zans Be'Eretz Hakodesh, 1980), p. 16. Paradoxically, Herzl too relates anti-Semitism to the prominence of Jewish financiers: see S. Avineri, *Herzl, The Matter of the Jews: Journals* (Jerusalem: Mossad Bialik and Hasifria Hazionit, 1997), vol. 1, p. 18.
25 L. Pinsker, *Auto-Emancipation* (New York: Zionist Organization of America, 1948).
26 E. Pawel, *The Labyrinth of Exile: A Life of Theodor Herzl* (New York: Farrar, Straus & Giroux, 1989) p. 197.
27 Oz Almog, *The Sabra: The Creation of the New Jew* (Berkeley: University of California Press, 2000).
28 It would be inaccurate to say that nothing at all has been done, but the continuing financial problems of the "Voice of Israel in Arabic" radio station and the poor intelligence resources given to the Israeli psychological warfare unit over the years show that the issue had a very low place on the Israel's list of the security priorities.
29 J. Ellul, *Propaganda: The Formation of Men's Attitudes* (New York: Vintage Books, 1973), p. 189.

30 The Israeli Foreign Ministry had published a series titled "Double Talk" for a decade beginning in the mid-1970s. These documents are available in the Foreign Ministry Archives, Jerusalem.
31 See the US Commission on International Religious Freedom press release of 6 November 2008, "USCIRF Confirms Material Inciting Violence, Intolerance Remains in Textbooks Used at Saudi Government's Islamic Saudi Academy", available at http://www.uscirf.gov/index.php?option=com_content&task=view&id=2206&Itemid=1.
32 The tide started with Prof. Benny Morris' book on the roots of the refugee problem and evolved into a fashionable academic trend called the "new historians" which was quickly embraced by the Palestinians and left-wing academics in the West. Morris realized the impact of revisionist conscripted history and altered his views, but then he was no longer useful for the circle he unwittingly helped create. See B. Morris, *The Birth of the Palestinian Problem, 1947–1949* (New York: Cambridge University Press, 1987).
33 See Joel Fishman, "Information Policy and National Identity: Israel's Ideological War", ACPR Policy Paper No. 142, available at http://www.acpr.org.il/pp/pp142-Fishman-E.pdf.
34 This is the view of the secular nationalist factions of the PLO such as Fatah. See R. Khalidi, *Palestinian Identity: The Construction of Modern National Consciousness* (New York: Columbia University Press, 1998), p. 128. On Bible events, see Hassan Haddad, "The Biblical Bases of Zionist Colonialism", *Journal of Palestine Studies* 3.4 (1974), pp. 97–113.
35 Prime Minister Ehud Barak said that if he were a Palestinian he would have joined one of the underground Palestinian organizations. See "Soldier Turned Politician Who Spent His Life Fighting Arabs", *The Guardian*, 28 December 2008, available at http://www.guardian.co.uk/world/2008/dec/28/ehud-barak-profile.
36 Among other things, one can see the reflection of this phenomenon in the names given to children born in the settlements. Birnbaum calls these names "theophoric". See B. Birnbaum, "Proper Names of Secular and Religious Children Born between 1983–1992", MA thesis, Bar-Ilan University, 2000.
37 See G. Aran, "From Religious Zionism to Zionist Religion: The Roots of Gush Emunim and its Culture", Ph.D. thesis, Hebrew University of Jerusalem, 1987, p. 20.
38 Even though even there is a growing trend towards adopting the "settler" ethos.
39 Mishna, Gittin 4:6.
40 The most famous example is that of the Maharam of Rotenburg, the leader of German Jewry in the fourteenth century, who died in jail

## NOTES

because he refused to allow his congregation to pay the ransom.

41 See "PM Netanyahu's Remarks at a Special Press Conference Regarding the Continued Efforts to Release Kidnapped Soldier Gilad Shalit", a speech delivered 1 July 2010, available at http://www.pmo.gov.il/PMOEng/Communication/PMSpeaks/speechshalit010710.htm.

42 According to the Central Bureau of Statistics, 2.5 million Israelis traveled abroad in 2009, a small decease in comparison to 2008, and the Israeli population in 2010 is slightly over seven million. See http://www.ynetnews.com/articles/0,7340,L-3833374,00.html.

43 This lobby is the Anti-Discrimination Committee (ADC), headed by Dr. James Zogby, a most eloquent and creative US citizen of Lebanese descent. In the past decade the organization was pushed aside to some extent by the far more sophisticated religious Islamic organization, the Council for American-Islamic Relations (CAIR).

44 Menachem Begin was the first to publicly use evangelical support during the Lebanese war of 1982.

45 N.S. Atik, *Justice and Only Justice: A Palestinian Theology of Liberation* (New York: Orbis, 1989). Also see Dr. Atik's official website http://www.sabeel.org/.

46 Even so, between Muslims and Jews there were also residues of hostility that in extreme cases reached persecution of Jews and in other cases caused continuing discrimination. In this context, one must reject as a distortion of history the Palestinian claim that until the Zionists came the relationship between Muslims and Jews was one of love and peace. Indeed, social developments in the West supplied additional confirmation of this history: one can see a return to anti-Semitism in the extreme right-wing political parties in Europe.

47 See Y. Manor, "Integrating Hasbara into Policy", *Jerusalem Quarterly* 33 (Fall, 1984), pp. 25–39.

48 See Linda Sandler, David Voreacos, and Edvard Pettersson, "Rabbi Fraud Case in Los Angeles Echoes N.J. Scandal", Bloomberg, 3 August 2009, available at http://www.bloomberg.com/apps/news?pid=newsarchive&sid=aK1.Euuzx0c8.

49 See http://www.hagshama.org.il/doingzionism/resources/view.asp?id=1631; this document was originally published as Herbert Parzen Herzl, *Herzl Speaks: His Mind on Issues, Events and Men* (New York: The Herzl Press, 1960).

50 Ben-Gurion allegedly said this during a government meeting in 1951 of which there is no transcript. His biographer Shabtai Tevet claimed Ben-Gurion never made such a statement; nevertheless, this saying is common in Israeli discourse.

## 5 Countering Islamic Terrorism: The Psychological Warfare Perspective

1 It took just one story in the *New York Times* to close down the Pentagon's Office of Strategic Influence. See Eric Schmitt and James Dao, "A 'Damaged' Information Office is Declared Closed by Rumsfeld", *New York Times*, 17 February 2002, http://www.nytimes.com/2002/02/27/international/27MILI.html.
2 The Chinese strategist Sun Tzu of the fifth century BC is often quoted in PSYOP literature. See *The Art of War*, translated by R. D. Sawyer (Boulder, Colorado: Westview Press, 1994).
3 T.A. Qualter, *Propaganda and Psychological Warfare* (New York: Random House, 1962), p. 55
4 P.M. Taylor, *British Propaganda in the Twentieth Century: Selling Democracy* (Edinburgh: Edinburgh University Press, 1999), pp. 35–43.
5 For a detailed description of the evolution of this terminology see P.M. Taylor's foreword to R. Cole, *International Encyclopedia of Propaganda* (Chicago: Fitzroy and Dearborn Publishers, 1998), pp. xix–xxiii. W.L. Hixson outlines the American fear of propaganda and the terminology it created ("public diplomacy" etc.) in his book *Parting the Curtain: Propaganda, Culture, and the Cold War, 1945–1961* (New York: St. Martin's Press, 1998).
6 George M. Foster, *Applied Anthropology* (Boston: Little, Brown, and Company, 1969), p. 187.
7 Surprisingly the solution was found by a non-Japanese speaking anthropologist. See C. Roetter, *Psychological Warfare* (London: B.T. Batsford, 1974), pp. 136–143.
8 E.T. Hall, *The Hidden Dimension* (Garden City, NY: Doubleday, 1966); and G. Hofstede, *Culture's Consequences: International Differences in Work-Related Values* (Beverly Hills, California: Sage Publications, 1980).
9 See Roetter, *Psychological Warfare*, pp. 136–143.
10 See A.B. Gilmore, *You Can't Fight Tanks with Bayonets: Psychological Warfare against the Japanese Army in the Southwest Pacific* (University of Nebraska Press, 1998), p. 6.
11 L. Farago, *German Psychological Warfare* (New York: Putnam & Sons, 1942).
12 See Guy Raz, "Simulated City Preps Marines for Reality in Iraq", broadcast on NPR on April 13, 2007, available at http://www.npr.org/templates/story/story.php?storyId=9573747.
13 US Air Force Doctrine Document 2-5.3, *Psychological Operations* (August, 1999), esp. pp. 28 and 43. The document is available on the internet at www.iwar.org.uk/PSYOPs/resources/us/afdd2-5-3.pdf.
14 A.M. Burton, *Urban Terrorism: Theory, Practice and Response* (London: Leo Cooper, 1975), pp. 220–226.

## NOTES

15 Capt. M.T. McEwen, "Psychological Operations against Terrorism: The Unused Weapon", *Military Review* 66.1 (January, 1986), p. 65, lists a few PSYOP techniques according to his contemporary manual, such as planting rumors of betrayal, sowing dissent, and describing the procedures to surrender.

16 In the 1970s M. Tugwell, a former MI officer, prepared for his PhD thesis a list of themes terrorists had used in their messages through history; it is just as relevant today as when it was written. See M. Tugwell, *Revolutionary Propaganda and Possible Counter-Measures*, unpublished PhD thesis, King's College, London, 1979. Other relevant insights by Tugwell can be found in his article "Terrorism and Propaganda: Problem and Response", in P. Wilkinson and A.M Stuart, eds., *Contemporary Research on Terrorism* (Aberdeen: Aberdeen University Press, 1987), pp. 409–418.

17 I. McDowell, *Reuters Handbook for Journalists* (Oxford: Butterworth-Heinemann, 1992), pp. 147–148.

18 Olaf Dilling, "Does 'Infinite Justice' Lead to Enduring War?" *German Law Journal* 2.16 (2001), p. 10.

19 As was recommended for military affairs as early as 1995 by F. Aukofer and W. Lawrence in their document *America's Team: The Odd Couple – A Report on the Relationship Between the Media and the Military* (Nashville, Tenn.: The First Amendment Center, 1995). This was one of the more important non-classified studies written during the 1990s that attempted to reconcile the conflicting interests of the military and the media. The next development was embedded journalism, but the media found this arrangement too restrictive and criticized the government after the war.

20 B. Jenkins, *International Terrorism* (Los Angeles: Crescent, 1975), p. 4.

21 See for example "Obama Must Pay Heed to al-Qaeda's Quest for Biological Weapons", *The Washington Post*, 3 February 2010.

22 The US had faced similar issues in the Pacific Theatre during World War II. During the Cold War, on the other hand, the US was again facing an enemy whose culture was largely familiar: see L. Bogart, *Premises for Propaganda: The US Information Agency's Operating Assumptions in the Cold War* (New York: Macmillan, 1976).

23 During the European airport closure as a result of the volcanic eruption in Iceland in 2010 a group of Saudi students told a reporter that the real reason behind the closure was Britain's wish to keep tourists spending more money in the UK. See *Haaretz*, 18 April 2010, available at www.haaretz.co.il.

24 M. Kramer, "The Status of Middle Eastern Studies in America", in *Peacewatch/Policywatch Anthology 2001: A Year of Terror* (Washington, DC: The Washington Institute, 2001). See also Martin

Kramer, *Ivory Towers on Sand: The Failure of Middle Eastern Studies in America* (Washington, DC: Washington Institute for Near East Policy, 2001).
25 Since these are first (and belated) steps, many difficulties have arisen regarding the loyalty of the candidates. See D. Eggen, "FBI Still Lacking Arabic Skills", *The Washington Post*, 11 October 2006, p. A01.
26 On Arab allegations about popular stereotypes see J.C. Shaheen, *The TV Arab* (Bowling Green, Ohio: Bowling Green University Popular Press, 1984).
27 See Shaheen, *The Arab TV*.
28 See Gilbert Sewall, *Islam in the Classroom: What the Textbooks Tell Us* (New York: American Textbook Council, 2008), available at http://www.historytextbooks.org/islamreport.pdf.
29 See "Guilty Verdicts in Holy Land Foundation Retrial", 24 November, 2008, available at http://cbs11tv.com/local/holy.land.retrial.2.872727.html.
30 The timely release of information about Al Qaeda experiments in biological warfare shows that the government has not dropped the PSYOP ball completely. See Judith Miller, "Qaeda Videos Seem to Show Chemical Tests", *New York Times*, 19 August 2002, Section A, p. 1.
31 In light of the evidence that many of the hijackers apparently were not aware of the fact they are going to die with the passengers.
32 See Somini Sengupta, "Death by Stoning Sentence Overturned: Nigerian Woman Acquitted of Adultery on Appeal", *New York Times*, 26 September 2003.
33 This type of message was delivered in Afghanistan during operation Enduring Justice. An aerial photo showed the license plate of the car of Omar, Bin Laden's second in command, with sniper crosshairs centered on it. The message is quite clear.
34 http://psywar.psyborg.co.uk/afghanistan.shtml.
35 Supplying information about soldiers who have been taken prisoner is one such technique, which was used for example by Egyptian PSYOP radio after the Yom Kippur War. In addition, a masterly use of this technique was made by Hezbollah in Lebanon through the TV station Al Manar. See Frederic M. Wehrey, "Clash of Wills: Hizballah's Psychological Campaign Against Israel in South Lebanon", *Small Wars & Insurgencies* http://www.informaworld.com/smpp/title~db=all~content=t713636778~tab=issueslist~branches=13-v1313.3 (2002), pp. 53–74.

## 6 Cultural Warfare: Secularization Defense Initiative

1 About diversity of meanings of cultural warfare see: Demerath, N. J. III (2005) "The Battle Over a U.S. Culture War: A Note on Inflated

# NOTES

Rhetoric Versus Inflamed Politics", *The Forum*, Vol. 3: Iss. 2 http://citeseerx.ist.psu.edu/viewdoc/download?doi=10.1.1.169.7179&rep=rep1&type=pdf. A recent work coming from US military thought is formulating culture as a major strategic tool, see (Ltc). S. A. Green (2008) (unpublished dissertation) "Cognitive Warfare", US Joint Military Intelligence College. http://www.scribd.com/doc/20459296/Stuart-Green-LTC-USN-Cognitive-Warfare-02-July-2008-Final.

2 See Raphael Israeli, *Islamikaze: Manifestations of Islamic Martyrology* (London: Frank Cass Publishers, 2003), pp. 444–445. Israeli, somewhat provocatively, suggests that radical Islam has launched a seven-stage battle plan. Having first put the West on the defensive, it then hopes to drive a wedge between the United States and its European allies by humiliating the United States in the global arena. This will allow it to focus its attention on Europe, where it will start with "freeing" historic al-Andalus (Spain and southern France), then going on to liquidate both Zionism and the Jews. Having dealt with Europe, it will turn to the task of destroying the United States. It will do all this by, among other things, mobilizing Muslim converts in the West, offering financial support to the families of suicide bombers, and launching a strident education campaign among the Islamic youth. See also the Al Qaeda Manual at http://disastercenter.com/terror/. Though it mentions the West, the manual mainly targets the secular Arab states. Israeli dismisses the idea that democracy can take root in the Arab states given that, so far, attempts to democratize them have failed signally, with the possible exception of Turkey, which is hardly a success story either. He nevertheless argues the case for reeducating those sections of Muslim society that have been infected by radical Islamic terrorism; though he neglects to address the question of how exactly to do this (Israeli, *Islamikaze*, p. 345). In central Asia, the Hizb ut Tahrir, which has a stranglehold on the region, hopes to resurrect the Caliphate and unite all Muslim lands, including China's Xing Jiang province, into a single state: see Ahmed Rashid, *The Rise of Militant Islam in Central Asia* (New Haven, Conn.: Yale University Press, 2002), pp. 115–136.

3 Samuel P. Huntington, *The Clash of Civilizations and the Remaking of World Order* (New York: Simon & Schuster, 1998).

4 http://www.airpower.maxwell.af.mil/airchronicles/cc/schwalbe3.html.

5 Natan Sharansky, *The Case for Democracy: The Power of Freedom to Overcome Tyranny and Terror* (New York: Public Affairs, 2004), pp. 5–11.

6 Address before the Joint Session of Congress on the State of the Union, January 25, 1994. http://www.washingtonpost.com/wp-srv/politics/special/states/docs/sou94.htm.

7 Dean V. Bakbst, "Elective Governments – A Force For Peace", *The Wisconsin Sociologist*, Vol. 3, No. 1 (1964), pp. 9–14. Even his critics agree with the basic facts and only dissent regarding his interpretation: see Melvin Small and David J. Singer, "The War Proneness of Democratic Regimes, 1816–1965", *Jerusalem Journal of International Relations* 1 (1976), pp. 50–69; Stuart A. Bremer, "Dangerous Dyads: Conditions Affecting the Likelihood of Interstate War, 1816–1965", *The Journal of Conflict Resolution*, Vol. 36, No. 2 (June, 1992), pp. 309–341; idem, "Democracy and Militarized Interstate Conflict, 1816–1965", *International Interactions*, Vol. 18, No. 3 (1993), pp. 231–249; Bruce Russett, *Grasping the Democratic Peace* (Princeton, NJ: Princeton University Press, 1993); Rudolph J. Rummel, *Power Kills: Democracy as a Method of Nonviolence* (New Brunswick, NJ: Transaction Books, 1997); Zeev Maoz, "The Controversy over the Democratic Peace: Rearguard Action or Cracks in the Wall?" *International Security*, Vol. 22, No. 1 (1997), pp. 162–198. Most of the attempts to criticize the factual thesis of the theory had to rely on countries that were not full-fledged democracies or young democracies or on conflicts that were lesser than wars. See, for example, Thomas Schwartz and Kiron K. Skinner, "The Myth of the Democratic Peace", *Orbis*, Vol. 46, No. 1 (1992), pp. 159–172.

8 Michael W. Doyle, *Ways of War and Peace* (New York: W.W. Norton & Co, 1997); Kevin Shimmin, "Critique of R. J. Rummel's 'Democratic Peace' Thesis", *Peace Magazine*, Vol. 15, No. 5 (1999), http://www.peacemagazine.org/archive/v15n5p06.htm; Christopher F. Gelpi and Michael Griesdorf, "Winners or Losers? Democracies in International Crisis, 1918–94", *American Political Science Review*, Vol. 95, No. 3 (2001), pp. 633–647, http://www.duke.edu/~gelpi/ democratic.winners.pdf; Gilat Levy and Ronny Razin, "It Takes Two: An Explanation for the Democratic Peace", *Journal of the European Economic Association*, Vol. 2, No.1 (2004), pp. 1–29, https://mitpress.mit.edu/journals/pdf/jeea_2_1_1_0.pdf; James Lee Ray, "Does Democracy Cause Peace?" *Annual Review of Political Science* 1 (1998), pp. 27–46, http://www.mtholyoke.edu/acad/intrel/ray.htm; idem, "Constructing Multivariate Analyses (of Dangerous Dyads)", *Conflict Management and Peace Science* 22 (2005), pp. 277–292, http://sitemason.vanderbilt.edu/files/ g/gDf5Ty/rayconstructingmultivaraite.pdf.

9 It is sometimes argued that it is precisely this weakness – the reluctance or inability to engage in long-term and bloody conflicts – that guarantees that democracies (unless attacked) will only take part in just wars: wars whose legitimacy is above and beyond question. It is the only thus that they can justify their actions morally, and so persuade their public to keep on fighting. Some go even further, insisting that this reluctance

is the root of the assertion that democracies are fundamentally peaceful. This spurious inference, lacking in all logic, is a perverse if understandable attempt to represent a negative as a positive. It is also plain wrong.

Leaving aside the question of whether there is such a thing as an absolute moral right, no state can allow its military authority, not to mention its existence, to be dependent upon the adoption of a wholly, 100 percent, just policy, let alone one that is just in the eyes of all, assuming such a consensus is even possible. Moreover, when at war, the public's sense of right or wrong is not necessarily rooted in questions of moral principle. Rather than the product of a just code of ethics, born of an attempt to wrestle with difficult philosophical conundrums, willingness to engage in war is usually the offspring of instinct and emotion fed by a constant diet of bloodcurdling images, inflammatory media, and political sound bites.

10 Charles de Secondat, Baron de Montesquieu, *The Spirit of the Laws*, trans. T. Nugent (New York: Macmillan, 1949), vol. 1, chap. 3, p. 3.
11 That was also the fate of the "New Middle East", a project to promote Arab economic development. See Shimon Peres, *The New Middle East* (New York: Henry Holt & Co., 1993).
12 Robin W. Winks and Joan Neuberger, *Europe and the Making of Modernity: 1815–1914* (Oxford and New York: Oxford University Press, 2005); Simon M. Dixon, *The Modernisation of Russia: 1676–1825* (Cambridge: Cambridge University Press, 1998); Samuel N. Eisenstadt, *Tradition, Change and Modernity* (New York: John Wiley & Sons, 1973).
13 Matthew Spink, *John Huss: A Biography* (Princeton, NJ: Princeton University Press, 1968), pp. 132–140.
14 Harold J. Grimm, *The Reformation Era, 1500–1650* (New York: Macmillan Publishers, 1973), pp. 89–97.
15 A. G. Dickens, *The English Reformation* (London: B. T. Batsford, 1964), pp. 137–145.
16 In PSYOP terms, this generally means the news.
17 Sigmund Freud, *Group Psychology and the Analysis of the Ego* (London: Hogarth Press, 1959), pp. 87–88.
18 S. Ettinger, *Between Russia and Poland* (Jerusalem: Zalman Shazar Center, 1994), pp. 323–324.
19 In Afghanistan, US PSYOP units distributed pictures of Taliban enforcers whipping and executing women in public, see http://psywarrior.com/Herbafghan.html.
20 In the preface to his book, *Al-Muhaddithat: The Women Scholars in Islam*, A. M. Nadwi states his intention to debunk the West's spurious image of Muslim women as oppressed, second-class citizens. The book centers on the part women played in the dissemination of the

Hadith, by both interpreting and teaching Islamic lore. See A. M. Nadwi, *Al-Muhaddithat: The Women Scholars in Islam* (Oxford: Interface Publication, 2007), pp. xi–xii.

21 See for example: Roksana Bahramitash, "Revolution, Islamization and Women's employment in Iran", The Brown Journal of World Affairs, Winter/Spring 2003 – Volume IX, Issue, pp. 229–241.

22 Quran, Sura 2:228: . . . Wives have the same rights as the husbands have on them in accordance with the generally known principles. Of course, men are a degree above them in status . . .

23 An example from the Ground Zero debate, D. Foster, The Slums of Rauf: http://www.nationalreview.com/corner/245321/slums-rauf-daniel-foster.

24 Both diplomats and state department officials are likely to object to any attack on the Arab and Muslim elites' probity, fearing that this might precipitate political unrest in countries currently allied to the West. While a valid point, the West ought nevertheless to measure its short-term interests, namely supporting friendly Arab regimes, against its long-term goals, that is, winning the war against radical Islam.

25 The Hassidic movement is occasionally accused of having accelerated the disintegration of traditional Jewish society. Even if this were the case, modernity was certainly not part of the movement's agenda, whose role was limited to undermining their community's existent forms of authority.

26 Irshad Manji, *The Trouble with Islam: A Muslim's Call for Reform in Her Faith* (New York: St. Martin's Press, 2004).

27 Holders of this view also argue that Islam could easily embrace democracy, as both share the same egalitarian values: democracy holds that all people are equal before the law, while Islam believes in the equality of men before God. See Noah Feldman, *After Jihad: America and the Struggle for Islamic Democracy* (New York: Farrar, Strauss and Giroux, 2003), p. 76.

28 For Shi'ite apologetics on this issue, see *The Shi'ite Encyclopedia of Islam*, chap. 7, at http://www.al-islam.org/encyclopedia.

29 There is likely to be some pressure, which should be resisted, to use these modes of influence for more conventional intelligence purposes.

30 Walter L. Hixson, *Parting the Curtain: Propaganda, Culture, and the Cold War, 1945–1961* (Basingstoke: Palgrave Macmillan, 1997).

31 On END see: http://www.spokesmanbooks.com/Spokesman/PDF/Coates100.pdf. The Soviets also bamboozled people into leaking stories to the press and set up front organizations along the lines laid down by the German communist Willy Muenzenberg in the 1920s. In 1949, for example, the Soviet Union founded the World Peace Organization; its aim was to support Stalin's initiative for world peace.

## NOTES

It is worth noting that several currently highly respected institutions began life as Soviet front organizations. In Soviet Strategy such operations were called "Active Measures and include operations such as alarming the British against US placed ballistic missiles that their country will be obliterated in case of a US–USSR nuclear confrontation.

Other operations include diverting foreign policy in India and manipulating wheat prices on the international market. See: http://intellit.muskingum.edu/russia_folder/pcw_era/sect_16a.htm
See John Boynton, *Aims and Means* (London: Dufour Editions, 1964); and Richard Deacon, *The Truth Twisters* (London: MacDonald, 1987). Not that the Americans were remiss in this respect: see Leo Bogart, *Premises of Propaganda: The U.S. Information Agency's Operating Assumptions in the Cold War* (New York: Macmillan, 1976). Following the lead of the Soviet model, radical Islam both exploits existing non-governmental organizations and sets up front organizations of its own. See http://ngo-monitor.org.

32 Such negligence is not exclusive to the Arab-Islamic world: Peter the Great's campaign to modernize and secularize seventeenth-century Russia failed for much the same reasons.
33 Japan's entry into the war in 1941 complicated matters, leading to some mostly improvised attempts to understand Japanese culture.
34 See Martin Kramer's critique of Edward Said's *Orientalism*, which Kramer claims has helped silence any dissenting voices in the field of Middle East studies. M. Kramer, *Ivory Towers on Sand: The Failure of Middle Eastern Studies in America* (Policy Papers, Washington Institute for Near East Policy, Washington, DC), no. 58. Ibn Warraq, *Defending the West: A Critique of Edward Said's Orientalism* (Amherst, NY: Prometheus Books, 2007.
35 Introducing modern economic and technological infrastructures before the hoped-for cultural transformation is complete may result in the misuse of those infrastructures, the greatest danger being that they will be employed to acquire new, advanced weapon systems. This pitfall can be overcome by making sure that these technologies are dedicated solely to improving daily life, and are not exploited for any other purpose.
36 The decision-making dynamic of the Bush administration is yet to be fully divulged. However, judging by the results of their decisions (and Bush's own public image), it appears that the administration's somewhat gung-ho policy-makers were unappreciative of academic theorizing. Academics, consultants and government agencies have created a "democratization industry" each for its own interests yet disregarding objective conditions in the Arab world. See: *Political Liberalization and Democratization in the Arab World: Theoretical*

*Perspectives*, edited by Rex Brynen, Bahgat Korany and Paul Noble (Boulder, CO: Lynne Rienner), p. 333.

# Index

9/11 terrorist attack, 3, 15, 31, 36, 107
  visual images, 113–14

Abbas, Abul, 52–3
Abu Ghraib affair, 12
Abu Mazen, 24
Abu Rahmeh, Talal, 61
*Achille Lauro* cruise ship, 52–3
ADC (Anti-Discrimination Committee), 164*n*
Afghanistan
  Bin Laden's presence, 104
  Dutch forces, 147*n*
  status of women, 110, 129, 170–1*n*
  Western image of Muslims, 110–11
Afghanistan War, 3, 14
  consolidation PSYOP, 148–9*n*
  "Enduring Freedom" operation, 102
  Special Forces deployment, 104
  suicide bombers, 1–2
  US PSYOP campaigns, 7–8, 9–11, 115, 129, 148–9*n*, 170–1*n*
  Western PSYOP campaigns, 7–8, 9–11, 99, 110–11, 115, 148–9*n*, 167*n*, 170–1*n*
Ahaseurus, 72
Ahmadiyah sect, 132
AIPAC (American Israel Public Affairs Committee), 90
Akiva, Rabbi, 74–5
*Al Arabia*, 2
Al Hurra (The Free), 12
*Al Jezeera* news network, 2, 11, 12
Al Muhajirun, 110
Al Najah University, 44
Al Qaeda
  biological warfare experiments, 106, 167*n*
  delivery channels, 115
  Manual, 168*n*
  religious motives, 113
  support of Muslims in West, 110
  threat to West, 104, 105

US PSYOP campaigns, 104, 105
Western PSYOP campaigns, 104, 105, 112
Al-Aqsa Intifada *see Intifada* (2000–2005)
Almog, Doron, 152*n*
American Israel Public Affairs Committee (AIPAC), 90
Amnesty International, 25
Annan, Kofi, 29, 153*n*
anthropological intelligence, 6, 15–16, 99–100, 158*n*
Anti-Discrimination Committee (ADC), 164*n*
anti-Semitism, 77, 78, 84
  accusations against Israel, 47
  extreme right-wing, 164*n*
  hasbara, 93–5
  Jewish financiers, 162*n*
Apion, 73, 161*n*
Arab anti-Semitism, 78, 84
Arab diaspora, 90
Arab states
  image projection to the West, 84
  information ministries, 42, 147*n*
  PSYOP campaigns, 147*n*
  targeted by Al Qaeda, 168*n*
  *see also* Gulf States; Iraq; Saudi Arabia
Arab world
  cultural complexity, 15–16, 107–8
  Israeli attitude, 49
  Israeli hasbara, 89
  personal honor (face), 58, 107
Arab–Israeli War (1948), 85
Arad, Ron, 88
Arafat, Suha, 55, 153*n*
Arafat, Yassir
  attitude to the truth, 58, 60
  Bush's refusal to negotiate with, 36
  death of, 67
  destruction of Zionist state, 42
  diplomatic efforts on the West, 67
  embracing of foreign media, 31, 65

# INDEX

guerilla warfare, 67
as head of Palestinian Authority, 53
importance of PSYOP, 42
Jenin refugee camp, 28
*Karin A* incident (2002), 54, 60
"Lebanonization" process, 54
lessening of tension, 56–7
Oslo Accords (1993), 43
overall goals of the Palestinians, 45
pro-Arafat biographies, 159$n$
PSYOP operational techniques, 42, 46, 50
speech at UN, 160$n$
steady political line maintenance, 70
Arikat, Saib, 42
Ashrawi, Hanan, 90
assets/liabilities, PSYOP operational techniques, 51, 81–2, 87
Atatürk, 141
Atik, Naim, 91, 156$n$
atrocity propaganda *see* demonization themes
attack and initiative, as PSYOP strategy, 53–4
Aukofer, F., 166$n$
Ayash, Yahyah, 48

*Baghdad Times*, 12
Barak, Ehud, 163$n$
Begin, Menachem, 164$n$
Beilin, Yossi, 94
Beit Lid, 48, 60
Ben-Gurion, David, 75, 78, 95, 165$n$
Betzelem, 159$n$
Bin Laden, Osama, 3, 14
  demonization theme, 112
  establishment of Al Qaeda, 104
  media management, 114–15
  US focus on, 111, 112
Bir Zeit University, 32–3, 44
black PSYOP, 8, 19–20, 148$n$, 149$n$
Britain
  black PSYOP, 149$n$
  cultural intelligence on Islamic world, 142
  domestic Muslim communities, 110
  overseas postings policy, 107–8
  PSYOP campaigns, 4
British School of Oriental and African Studies (SOAS), 108

Bush, George W.
  Bush–Sharansky doctrine, 118, 120, 121, 122, 140–1, 144–5
  decision-making dynamics, 172$n$
  democracies as peaceful, 120
  refusal to negotiate with Arafat, 36
  support for CAIR, 111

CAIR (Council for American-Islamic Relations), 111, 164$n$
Camden (London) Abu Dis Friendship Association, 54
capitalism, 123
Caterpillar Inc., 47, 154$n$
Catholic church, 126–7, 131
Chamor, 71
Christian Evangelicals, 90–1
Christianity
  cooperation with Karaites, 161–2$n$
  debate with Jews during the Middle Ages, 75
  Mendelsohn's views, 76
  modernization in Europe, 124–5, 126–7, 128–9, 131, 135–6, 139
  relations with Islam, 109
clergy, anti-clerical offensive, 126–8, 131, 142
Clinton, Bill, 120
Clinton, Hilary, 55, 153$n$
CNN, 114–15
Cold War, 1, 2
  black PSYOP, 149$n$
  cultural intelligence, 141
  liberation theology, 91
  lifestyle promotions, 134
  PSYOP campaigns, 4–5
  Sovietology, 108
communism, 1, 81, 83, 84, 111, 139
  *see also* Soviet Union
computerized PSYOP systems, 17–18
Condon, Jeanmarie, 154$n$
conscription themes, 109–10
consolidation PSYOP, 9–10, 11–12, 17, 101, 148–9$n$
Corrie, Rachel, 33, 154$n$
Council for American-Islamic Relations (CAIR), 111, 164$n$
counterterrorism PSYOP, 98–116
  advantages of, 101
  cultural intelligence, 106–8
  delivery channels, 103, 114–15

INDEX

counterterrorism PSYOP *(continued)*
　formulation of messages, 102, 108–9
　media management, 103, 105, 113–14, 115
　preparations, 104–6
　target audiences, 101–2, 109–12
　themes for enemy audience, 111–12, 114–15
　themes for home audience, 109–11
　themes for neutral audience, 112
　use of visual images, 113–14
cruelty, PSYOP operational techniques, 49–51
cultural intelligence, 15–16, 17, 99–101, 106–8, 141–2
cultural warfare, 117–46
　anti-clerical offensive, 126–8, 131
　cultivating the habit of critical thinking, 135–8
　freedom and the "Good Life", 132–5
　limits and qualifications, 141–4
　modernization and secularization of Muslim world, 118–19, 121–5, 127–8, 129–31, 133–5, 136–8, 139–46
　promoting internal religious reform, 131–2
　propagating dissident movements and secular ideologies, 139–41
　targeting the poor and dispossessed, 128–31
　weakness problem, 119–21

Dachoach-Halevi, Yonatan, 159$n$
Dead Sea Ahava products, 47
Def, Muhammad, 60
democracy
　Bush–Sharansky doctrine, 118, 120, 121, 122, 140–1, 144–5
　cultural warfare doctrine, 118–19
　engagement in wars, 120–1, 170$n$
　introduction to Islamic world, 5, 118, 120, 121–5, 139–41, 144–5
　and Islam, 171$n$
　just wars, 170$n$
　principal aim of, 120
　victory of, 139
　weakness of, 119–21, 145
"democratic peace theory", 120
democratic states
　media supervision, 64, 103

PSYOP campaigns, 5, 7, 98, 103
　reaction times, 111–12, 114
demonization themes, Palestinians, 24, 28–30, 32, 50, 55, 61, 71, 151$n$
determination, PSYOP operational techniques, 49–51
Dina, 71
diplomatic networks, *Intifada* (2000–2005), 32
al-Dura, Muhammad, 33–4, 61, 63, 92
Durban Conference (2001), 47

Eban, Abba, 90
Eliezer, R., 74
Eliyahu ben Amozegh, Rabbi, 76
Ellul, J., 82
Enlightenment, 120, 124, 139
equality, principle of, 128
Etzel underground organization, 79
European Nuclear Disarmament(END), 137

Farago, Ladislas, 99–100
fascism, 139
*Fatwa* (religious edict), 134
film newsreels, as PSYOP medium, 13
Fishman, Joel, 85
FRANCE 2 television station, 34, 61
France, modernization, 124
French Revolution, 124
Freud, Sigmund, 128
fundamentalist Islam
　cultural warfare, 132, 136
　familiarity with West, 2–3
　influence on radical Islamic organizations, 1
　PSYOP campaigns, 3, 5, 14
　recruitment in the West, 149$n$
　threat to West, 4, 5
　Western PSYOP campaigns, 4, 5, 19–20, 21, 110
funeral rituals, Palestinians, 55–6

Gadahn, Yahya, 147$n$
Galei Tzahal Radio Station, 42
Gaza, Israeli retreat, 22, 38
Gaza War (2009)
　Goldstone Report, 24, 45
　Operation Cast Lead, 39, 42, 44, 50, 158$n$
Gelber, Yoav, 155$n$
George, Alexander, 148$n$

Goebbels, Joseph, 59, 98
Goldstone Report, 24, 45
Goldwasser, Ehud, 66
Grossfeld, Stan, 159$n$
guilt, PSYOP operational techniques, 47, 82
Gulf States
  wealth gap, 131
  Western influence, 123
Gulf War (1990–91)
  importance of visual images, 9, 105, 113
  PSYOP campaigns, 8–9, 10, 13, 99, 100, 149$n$
Gulf War (2003), 13, 99, 100–1
Gush Katif, 57

el-Hadi, Mahedi Abed, 65
Hadith, 19, 171$n$
Hall, E.T., 100
Haman, 72
Hamas
  attitude to the truth, 60
  demonization themes, 55
  government in Gaza, 22, 38, 50
  guerilla warfare, 67
  lessening of tension, 57
  overall goals of, 45
  PSYOP strategies, 38, 42, 43, 45, 50, 52, 63
  regarded as intransigent, 53
  relations with Palestinian Authority, 37, 67
  revenge terror attacks, 48–9
  suicide bombers, 48, 49, 158$n$
  Tahadiyeh (temporary truce), 39, 52, 158
  treatment of collaborators, 50–1
Haniyeh, Ismail, 43, 45
hasbara
  and anti-Semitism, 93–5
  Israel, 70–1, 79–81, 95–7
  Jewish attitude, 71–7
  scale of attitudes, 93
  weaknesses of, 89–92
  and Zionism, 77–81, 86–7, 95
Hassidic movement, 171$n$
Hatib, Ghassan, 31, 65
Henry VIII, King of England, 126–7
Herzl, Theodor, 77, 78, 95
Hezbollah, 88, 102, 133
Hirsch, Rabbi Samson Raphael, 77

Hizbut Tahrir, 168$n$
Hofstede, Geert, 158–9$n$
Holy Land Foundation, 111
honor killings, 130
human rights
  importance in the West, 3
  liberal democracies, 145
  Palestinian focus on, 25–7, 38, 50
Huntington, Samuel P., 117, 120
Hurndall, Tom, 33
Hus, Jan, 126
Husseini, Faisal, 85

imperialism, 6, 122, 142
Information Warfare (infowar), 99
initiative and attack, as PSYOP strategy, 53–4
International Court in The Hague, 26
international law, *Intifada* (2000–2005), 24, 25, 26, 38
International Solidarity Movement, 33, 34
internet
  black PSYOP, 20
  cultural warfare, 135, 138
  *Intifada* (2000–2005), 32–3
  Iranian protests (2009), 125
  as PSYOP medium, 13–14, 43–4, 91, 106
  radical Islam, 2, 135, 143
  terrorist PSYOP, 115
*Intifada* (1987–91)
  civilian casualties, 47
  damning of Israel as a Nazi State, 27
  establishment of JMCC, 31
  Israeli motivation crisis, 87
  as non-violent campaign, 28, 32
  Palestinian PSYOP strategies, 40, 47, 62–3
  PSYOP delivery channels, 13, 20, 43
  PSYOP targeting of Israeli audience, 47
*Intifada* (2000–2005)
  aid workers, 35
  blackening of Israel's name, 24–8
  damning of Israel as a Nazi State, 25, 27
  foreign activists recruitment, 33
  human rights theme, 25–7, 38

# INDEX

*Intifada* (2000–2005) *(continued)*
  international law theme, 24, 25, 26, 38
  internet campaign, 32–3
  Israeli targeted assassinations policy, 25, 48, 50, 58
  Jenin Massacre, 28–30, 33, 34
  manufacturing incidents, 33–4
  military checkpoints, 25–6
  neutral audience, 22, 23, 24, 26, 33, 34, 35–6, 37
  nuclear weapons accusations, 25, 151*n*
  Palestinian demonization themes, 24, 28–30, 32, 151*n*
  Palestinian denial of responsibility, 27–8
  Palestinian diplomatic networks, 32
  Palestinian objectives, 24
  Palestinian PSYOP campaign, 22–38, 46, 50, 52, 67, 150*n*
  Palestinian use of media, 30–1, 36
  personal experience technique, 34–5
  persuasion techniques, 33–5
  "propaganda by deed" concept, 32
  PSYOP delivery channels, 20, 30–3
  responsibility evasions, 59
  security fence, 26–7, 62, 152*n*
  Siege of the Church of the Nativity, 28, 30, 92
  suicide bombers, 27–8, 32, 36, 158*n*
Iran
  educated population, 136
  hangings of homosexuals, 110–11
  *Karin A* incident (2002), 60
  nuclearization policy, 111
  protests (2009), 125
  PSYOP campaigns, 68
  "White Revolution", 141
  women's rights, 130
Iraq
  consolidation PSYOP, 17, 101, 149*n*
  cultural complexities, 15
  executions, 1, 2
  insurgents, 11, 12, 17
  invasion (2003), 54
  kidnappings, 147*n*
  Western PSYOP campaigns, 7–8, 11–13
  *see also* Gulf War (1990–91); Gulf War (2003)

Islam
  attitude regarding sex, 133
  care of the poor, 130–1
  as central to both state and society, 1
  and democracy, 171*n*
  diversity of, 132
  ideologies, 122
  influence on the West, 138
  relations with Christianity, 109
  status of women, 129–30, 171*n*
  Western analysis, 6
  Western converts, 2–3, 147*n*
  *see also* fundamentalist Islam; radical Islam
Islamic Jihad, 37, 53, 158*n*
Islamic world
  anti-clerical offensive, 127–8, 131
  Bush–Sharansky doctrine, 118, 120, 121, 122, 140–1, 144–5
  cultural complexity, 15–16, 107–8
  cultural warfare doctrine, 118–19, 121–5, 127–8, 129–31, 133–5, 136–8, 139–46
  democracy as by-product of capitalism, 122–3
  elites, 123, 131, 136–7, 138, 171*n*
  freedom and the "Good Life", 133–5
  habit of critical thinking, 136–8
  introduction of democracy, 5, 118, 120, 121–5, 139–41, 144–5
  modernization and secularization, 118–19, 121–5, 127–8, 129–31, 133–5, 136–8, 139–46
  stance on sex, 133
  status of women, 129–30
  targeting the poor and dispossessed, 128, 129–31
  US research on, 6, 108
  Western cultural intelligence, 15, 107–8, 141–2
  Western views of, 6
  *see also* Afghanistan; Gulf States; Iran; Iraq; Saudi Arabia
Islamization, 1
Isma'iliyah sect, 132
Israel
  accusations of anti-Semitism, 47
  assimilated Western-liberals, 94
  attitude to Arab world, 49
  black PSYOP, 20

# INDEX

broadcasting infrastructure, 70, 161*n*
contacts with world Jewry, 89–90
cultural intelligence, 16
damned as Nazi state, 25, 27
as a democracy, 70–1
Hasbara Ministry, 42
independence, 78
Intelligence Warfare Branch, 42
Israeli lack of historical knowledge, 85
Israeli liberals, 94
Israeli longing to go abroad, 89
Jewish settlers, 86–7
"longing for love" phenomenon, 78, 89, 91
*Malat*, 42, 43
moderate right wing, 94–5
national image, 70, 89
New Historians, 85, 163*n*
Operation Defensive Shield, 28, 158*n*
Palestinian blackening of Israel's name, 24–8
Palestinian demonization, 24, 28–30, 32, 50, 55, 61, 71, 151*n*
POWs and soldiers missing in action, 88
PSYOP (hasbara) campaigns, 70–1, 77–81, 86–7, 89–92, 93–7
public opinion, 82
relations with United States, 89–91
retreat from Gaza, 22, 38
socio-cultural developments, 82–3
suicide bombers, 107
Tahadiyeh (temporary truce), 39, 52, 158
target of Palestinian PSYOP campaigns, 8, 37–8, 39, 40, 44, 45–68
targeted assassinations policy, 25, 48, 50, 58
ultra-Orthodox Jews, 80, 83, 87, 93–4
War of Independence (1948), 85
*see also Intifada* (1987–91); *Intifada* (2000–2005); Lebanon War (1982); Oslo Accords
Israel Defense Forces (IDF)
cooperation with the Palestinians, 52
Galei Tzahal Radio Station, 42
*Intifada* (2000–2005), 25–6, 29, 30, 36
Jenin Massacre, 29
military education system, 87
missing soldiers unit, 88
motivation crises, 87
religious component, 87
Siege of the Church of the Nativity, 30
War of Independence (1948), 85
Zionist hasbara, 79
Israeli Ministry of Foreign Affairs
materials published, 91
Palestinian prisoners, 44
Ramallah lynching (2002), 91–2
sensitivity to anti-Israel sentiment, 89
Siege of the Church of the Nativity, 92
Israeli Ministry of Information and Diaspora Affairs, 42
Israeli, Raphael, 168*n*
Israeli-Palestinian conflict, main events, 68–9
Italy, kidnappings in Iraq, 147*n*

J-BIG (Jews for Boycotting Israeli Goods), 54
Japan
kidnappings in Iraq, 147*n*
World War II, 99, 100, 149*n*, 172*n*
Jenin Massacre, 28–30, 33, 34
Jenkins, B., 105
Jerusalem Media and Communication Centre (JMCC), 31, 65
Jewish community
modernization in Europe, 124–5, 129, 131–2, 135–6, 139, 171*n*
*see also* anti-Semitism
Jewish hasbara *see* hasbara
Jewish identity, 76, 86, 92–3
Jewish reform movement, 131
Jewish-Muslim relations, 92, 164*n*
Jews for Boycotting Israeli Goods (J-BIG), 54
el-Jibli, Razi, 56
Jibril, Ahmed, 52
"Jibril Deal" (1985), 88
*jihad*, 1, 102
JMCC (Jerusalem Media and Communication Centre), 31, 65
John XXIII, Pope, 126

**179**

# INDEX

*Johnny Comes Marching Home*, 87–8
Josephus Flavius, 73, 75
Judaism, 72–7, 94
justice, as PSYOP theme, 82

Karaites, 75, 161–2n
*Karin A* incident (2002), 54, 60
Karsh, Efraim, 159n
Kfar Darom, 56
Khomeini, Ayatollah, 130–1
kidnappings, Iraq, 147n
Klinghoffer, Leon, 53
Korean War, 100, 106
Kosovo, 13

Larssen, Terja, 29
Lavater, 75–6
Lawrence, W., 166n
Lebanon, Shi'ites, 88
Lebanon War (1982), 13, 67, 86, 87, 161n
Lehi underground organization, 79
Leo X, Pope, 126
Levi, 71
liabilities/assets, PSYOP operational techniques, 51, 81–2, 87
liberalism, 139
liberation movements
 PSYOP campaigns, 23
 *see also* Palestine Liberation Organization (PLO)
linkage, PSYOP operational techniques, 48–9
*Lusitania* medal, 106
Luther, Martin, 126

McEwen, M.T., 166n
Machiavelli, Niccolò, 146
Madrid bomb attack, 147n
Madrid Conference (1991), 67
Maharam of Rotenburg, 164n
Maimonides, 75
*Malat*, 42, 43
Manji, Irshad, 132
Manor, Yochanan, 93
marriage, forced, 130
marriage, short-term, 133
martyrdom, Muslim youths, 129
Marxism, 83–4, 123, 139
media
 Arafat's use of, 31, 65
 Bin Laden's use of, 114–15

counterterrorist PSYOP, 103, 105, 113–14, 115
*Intifada* (1987–91), 62–3
*Intifada* (2000–2005), 30–1, 36
Israeli, 65, 82
Palestinian PSYOP, 40, 64–5
radical Islamic groups, 2, 3
relations with US in First Gulf War (1990–91), 9
semantics, 102, 113
Soviet attempts to influence, 137
supervision in democracies, 64, 103
terrorist PSYOP, 103, 105
US government contacts, 115
*see also Al Jezeera* news network; internet; radio broadcasts; television news
Mendelsohn, Moses, 75–6
Middle Ages, 75, 138
Midrash, 73
*minim*, 73, 74, 161n
Mishna, 88
Mnaseas of Patra, 161n
modernization, 121–46
 Christian community in Europe, 124–5, 126–7, 128–9, 131, 135–6, 139
 Islamic world, 118–19, 121–5, 127–8, 129–31, 133–5, 136–8, 139–46
 Jewish community in Europe, 124–5, 129, 131–2, 135–6, 139, 171n
Molière, 127
Montesquieu, Charles de Secondat, Baron de, 121, 122
moral superiority theme, 112
Morris, Benny, 163n
Moses, 72
Muenzenberg, Willy, 172n
*Mujahideen*, 102, 113
mullahs, 2, 19
Muslim–Jewish relations, 92, 164n

Nadwi, A.M., 171n
Nasrallah, Hassan, 66
nationalism, 139
Nazis
 black PSYOP, 149n
 cultural knowledge, 100
Netanyahu, Binyamin, 88, 90
Netherlands, forces in Afghanistan, 147n

# INDEX

new leaf syndrome, PSYOP operational techniques, 52–3
newspapers, as PSYOP medium, 13

Obama, Barack, 144, 146
Operation Cast Lead, 39, 42, 44, 50, 158$n$
Operation Defensive Shield, 28, 158$n$
Orient House, 85
Orientalism, 142
Oslo Accords (1993), 39, 43, 45, 57, 67
ostracism, PSYOP operational techniques, 47–8

pacifism, 47, 87
Palestine Liberation Organization (PLO)
 diplomatic networks, 32
 doubletalk strategy, 58
 media supervision, 64–5
 PSYOP strategies, 81
 *see also* Oslo Accords (1993)
Palestinian Authority
 attention to history and documentation, 85
 basic goal of, 41
 denial of responsibility, 27–8
 diplomatic missions, 32
 information control, 37
 Jenin Massacre, 29
 as legitimate partner for Israel, 53
 lessening of tension, 56–7
 media supervision, 64
 Minister of Information, 42
 personal experience technique, 35
 prisoner releases, 57
 PSYOP courses, 43
 PSYOP delivery channels, 30
 relations with Hamas, 37, 67
 suicide bombers, 56
 Western audiences, 35
Palestinian Center for Policy and Survey Research, 65
Palestinian determination theme, 46, 49–51
Palestinian diaspora, 44
Palestinian National Bureau of Statistics, 159–60$n$
Palestinian–Israeli conflict, main events, 68–9

Palestinians
 black PSYOP, 20
 Christian world in the US, 91
 civilian casualties, 47, 58
 communist propaganda techniques, 83–4
 demonization themes, 24, 28–30, 32, 50, 55, 61, 71, 151$n$
 funeral rituals, 55–6
 human rights focus, 25–7, 38, 50
 ideological influences, 81
 Israeli response to PSYOP campaigns, 8
 Jewish settlers, 86–7
 "multi-messaging" techniques, 83–6
 opportunity creation, 61–2
 prisoner releases, 57
 PSYOP aimed at enemy audience, 37–8, 39, 40, 44, 45–61, 81–92
 PSYOP aimed at home audiences, 37, 40, 43, 45, 47, 48, 67, 151$n$
 PSYOP aimed at neutral audiences, 22, 23, 24, 26, 33, 34, 35–6, 37, 44–5, 60–1, 82
 PSYOP delivery channels, 30–3, 41, 43–4
 PSYOP operational techniques, 46–53, 61–5
 PSYOP organizational arm, 42
 PSYOP overall goals, 44–6
 PSYOP strategies, 53–61
 PSYOP themes, 24–30, 81–2, 151$n$
 PSYOP use of emotions, 62–3
 responsibility evasions, 59
 targeting of American Jewry, 51, 157$n$
 use of world Palestinian community, 90
 *see also Intifada* (1987–91); *Intifada* (2000–2005)
Palmach, 90
PASSIA, 65
Pearl, Daniel, 103
Pearleman, Adam, 147$n$
Peres, Shimon, 94
Peter the Great, 172$n$
Philo, 72
Pinsker, Leon, 77
Poland, kidnappings in Iraq, 147$n$
Ponsonby, A., 4
the poor, cultural warfare, 130–1

# INDEX

Protestantism, 131
PSYOP (psychological operation)
  aims, 5–6, 23, 122
  Arab states, 147$n$
  assessing effectiveness of, 65–6
  asymmetric conflicts, 23–4, 40–1
  audiences, 2, 4, 5–6, 10, 15–16, 22, 23, 24, 26, 33, 34, 35–6, 37–8, 39, 40, 43, 44–61, 67, 81–92, 96, 101–2, 109–12, 114–15, 151$n$
  computerized systems, 17–18
  consolidation campaign, 9–10, 11–12, 17, 148–9$n$
  cultural intelligence, 15–16, 17, 99–101, 106–8, 141–2
  defined, 40–2, 98
  delivery channels, 13–14, 30–3, 41, 43–4, 98–9, 103, 114–15
  enemy audiences, 5–6, 37–8, 39, 40, 44, 45–61, 81–92, 96, 111–12, 114–15, 151$n$
  fundamentalist Islam, 3, 5, 14
  home audiences, 23, 37, 40, 43, 45, 47, 48, 67, 109–11, 151$n$
  infrastructure requirements, 6
  limitations, 5–6
  moral factors, 5
  neutral audiences, 22, 23, 24, 26, 33, 34, 35–6, 37, 40, 44–5, 60–1, 82, 112
  news bulletins, 11
  operational techniques, 42, 46–53, 61–5, 81–2, 87
  radical Islamic organizations, 2, 4, 7, 20
  strategic level, 15, 42, 53, 117
  tactical level, 42, 53, 117
  technology, 13–15
  use of leaflets, 8–9, 10, 20, 43, 115
  versus negotiations, 41
  Western campaigns, 4–5, 7–8, 9–13, 19–20, 21, 99, 104, 105, 110–11, 112, 115, 148–9$n$, 167$n$, 170–1$n$
  Western moral aversion, 4
  see also black PSYOP; counter-terrorism PSYOP; cultural warfare; hasbara; terrorist PSYOP
public opinion
  Israeli hasbara, 81, 82, 89, 96
  negativity about Israel, 89
  Palestinian diplomatic network, 32
  Palestinian PSYOP campaigns, 26, 28, 30, 32, 82
  strategic PSYOP, 117

Qu'ran, 19

Rabin, Yitzchak, 159$n$
radical Islam
  appeal to Muslim youths, 129
  Arab minority support, 110
  black PSYOP, 20
  care of the poor, 130–1
  cultural intelligence, 15–16, 107–8
  cultural warfare doctrine, 117, 118–19, 121–5, 127–8, 141, 142–4, 146
  front organizations, 172$n$
  goal of global domination, 119
  influence over Muslims, 144
  PSYOP campaigns, 2, 4, 7, 20
  religious leaders, 127–8
  ruthless determination, 143–4
  status of women, 129–30
  suicide bombers, 18–19
  technical expertise, 14
  threat to the West, 1–8, 20–1, 119–21, 168$n$
  use of internet, 2, 135, 143
  Western counterterrorism PSYOP, 98–116
radio broadcasts
  black PSYOP, 20, 149$n$
  as PSYOP medium, 8, 10, 11, 13, 43
Radio Sawa (Friendship), 12
Ramallah lynching (2002), 91–2
Ravshake, 72
Reagan, Ronald, 4, 100
Regev, Eldad, 66
religion
  anti-clerical offensive, 126–8, 131
  challenges to, 135–6
  promoting internal religious reform, 131–2
Renaissance Europe, 124, 128
resoluteness theme, 112, 114
Reuters, 102
Rome, ancient, 1, 74
Rushdie, Salman, 112

Saadia Gaon, Rabbi, 161$n$, 162$n$
Saddam Hussein, 8, 9, 12, 113
el-Sahhaf, Muhammad Saeed, 54

# INDEX

Said, Edward, 6, 90, 108
satellite technology, as PSYOP medium, 13, 14
Saudi Arabia
    research funding, 6
    schools, 85
    wealth gap, 131
    Western influence, 123
secularization, 121–46
    Christian community in Europe, 124–5, 126–7, 128–9, 131, 135–6, 139
    Islamic world, 118–19, 121–5, 127–8, 129–31, 133–5, 136–8, 139–46
    Jewish community in Europe, 124–5, 129, 131–2, 135–6, 139, 171$n$
semantics, 102, 113
Sentient World Simulation (SWS), 149$n$
sexual relations, 132–3
Shah of Iran, 141
Shalit, Gilad, 44, 103
Sharansky, Natan, 118, 120, 121, 122, 140–1, 144–5
Sharet, Moshe, 78
Shari'a code, 1
Sharon, Ariel, 25, 27, 150$n$, 151$n$
Shechem, 71
Shehade, Salah, 47
Shi'ites, 133
Shik, Rabbi Moshe, 162$n$
Shikaki, Khalil, 65
Shimon, 71
Shimon Bar Yochai, Rabbi, 74
Shneersohn, R. Menachem Mendel, 77
smart phones, as PSYOP medium, 14
SOAS (British School of Oriental and African Studies), 108
Somalia, 111
"Sons of Noah" movement, 76
Soviet Union
    black PSYOP, 149$n$
    collapse of, 120
    communist propaganda techniques, 83–4
    front organizations, 172$n$
    influencing of Western media, 137
    liberation theology, 91

PSYOP campaigns, 4, 67, 147$n$
    see also Cold War
Spain, evacuation from Iraq, 147$n$
Stern, Menachem, 72
strength, inflated, PSYOP operational techniques, 51–2
Sudanese slave trade, 112
Sufism, 132
suicide bombers
    Afghanistan, 1–2
    Hamas, 48, 49, 158$n$
    *Intifada* (2000–2005), 27–8, 32, 36, 158$n$
    Palestinian strategy, 27–8, 32, 36, 46, 48, 56
    radical Islamists, 18–19
    West's ignorance, 107
Sun Tzu, 165$n$
surveys, 15–16, 65, 159–60$n$
SWS (Sentient World Simulation), 149$n$
Syria, 158$n$

Tahadiyeh (temporary truce), 39, 52, 158
Taliban, 3, 10, 148–9$n$
Talmud, 73, 74
Talmudic sages, 72, 73–4, 75
technology, PSYOP, 13–15
television, as PSYOP medium, 13, 43, 91
television news, 106, 114–15
    see also *Al Jezeera* news network
Temple Mount, 62
terrorism
    against Israel, 26, 36, 60
    Iraq, 12
    Operation Defensive Shield, 28
    PLO transformation, 32
    PSYOP campaigns, 40
    sporadic use of, 122
    threat to West, 2, 3–4, 21
    US intelligence, 7
    see also 9/11 terrorist attack; Al Qaeda; counterterrorism PSYOP; Taliban
terrorist PSYOP
    Al Qaeda threat, 104, 105
    delivery channels, 103, 114–15
    formulation of messages, 102
    target audience, 101–2
Tetzel, Johann, 126

# INDEX

Tevet, Shabtai, 165$n$
Third World, wars of independence, 50
Torah, 74, 75, 77, 162$n$
truth, as PSYOP strategy, 57–61
Tugwell, M., 102, 166$n$
Turkey, democratization, 168$n$
Twitter, 125

ultra-Orthodox Jews, 80, 83, 87, 93–4
underdog theme, 79, 103, 112
United Nations, "Zionism is Racism" resolution, 47
United States
  Al Qaeda threat, 104, 105
  appeasement mindset, 144
  black PSYOP, 8, 149$n$
  Christian Evangelicals, 90–1
  consolidation PSYOP, 148–9$n$
  cultural intelligence on Islamic world, 15, 107–8, 142
  cultural warfare doctrine, 129, 137, 142, 146
  delivery channels, 114–15
  domestic Muslim communities, 110
  efforts against terrorists, 3
  First Gulf War (1990–91), 8–9, 149$n$
  ignorance about suicide bombers, 107
  image of Muslims, 110–11
  intelligence organizations, 105
  Jewish community, 51, 89–90, 91, 157$n$
  media relations, 115
  Office for Strategic Deception, 7
  pluralist outlook, 144, 146
  post-World War II European programs, 137
  PSYOP campaigns, 4–5, 7–13, 100–1, 104, 105, 109–12, 148$n$, 149$n$
  PSYOP campaigns in Afghanistan, 7–8, 9–11, 115, 129, 148–9$n$, 170–1$n$
  PSYOP themes for enemy audience, 111–12, 115
  PSYOP themes for home audience, 109–11
  radical Islamic organizations threat, 2, 168$n$
  relations with Israel, 89–91
  research on Arab/Islamic world, 6, 108
  World War I, 106
  *see also* 9/11 terrorist attack; Cold War; counterterrorism PSYOP; Korean War; Vietnam War; the West

video cameras, as PSYOP medium, 43
Vietnam War
  black PSYOP, 149$n$
  consolidation PSYOP, 148$n$
  cultural intelligence, 100, 106
  PSYOP delivery channels, 115
  PSYOP messages by television, 13
  US military failure, 111
  Vietcong use of PSYOP, 4
Voice of America, 115
The Voice of Free Baghdad, 8
"Voice of Israel in Arabic" radio station, 163$n$

web sites *see* internet
Weinberger, Caspar, 4–5
Weizmann, Chaim, 78
the West
  Al Qaeda threat, 104, 105
  cultural intelligence on Islamic world, 15, 107–8, 141–2
  domestic Muslim communities, 110
  efforts against terrorists, 3
  fundamentalist Islam familiarity with, 2–3
  historic debt to Islam, 138
  ignorance about suicide bombers, 107
  image of Arabs, 84
  image of Muslims, 110–11
  importance of human rights, 3
  influence on Arab elite, 123
  introduction of democracy to Islamic world, 5, 118, 120, 121–5, 139–41, 144–5
  moral aversion to PSYOP, 4
  perception of wars of independence, 50
  PSYOP campaigns, 4–5, 7–8, 9–13, 19–20, 21, 99, 104, 105, 110–11, 112, 115, 148–9$n$, 167$n$, 170–1$n$
  radical Islamic organizations threat, 1–8, 20–1, 119–21, 168$n$
  views of Arab/Islamic world, 6

*see also* counterterrorism PSYOP;
    cultural warfare; United States
West Bank, security fence, 26–7, 62,
    152*n*
Western Wall Tunnels, 61–2
women's rights, 129–30
World Peace Organization, 172*n*
World War I
    PSYOP campaigns, 4, 13, 62, 98–9,
        141
    US involvement, 106
World War II
    black PSYOP, 149*n*
    PSYOP campaigns, 4, 13, 100, 141,
        149*n*

World Zionist Organization, 78
Wye Agreements (1988), 45

Yaakov, 71–2
Yaalon, Moshe, 60
Yassin, Ahmed, 57, 63
Yegar, Moshe, 161*n*
Yehuda, Rabbi, 74
Yishuv, 81

Ze'evi, Rechavam, 60
Zionism
    birth of, 139
    and hasbara, 77–81, 86–7, 95
Zogby, James, 164*n*

www.ingramcontent.com/pod-product-compliance
Lightning Source LLC
Chambersburg PA
CBHW071410300426
44114CB00016B/2255